SpringerBriefs in Earth System Sciences

South America and the Southern Hemisphere

Series Editors

Gerrit Lohmann
Lawrence A. Mysak
Justus Notholt
Jorge Rabassa
Vikram Unnithan

For further volumes:
http://www.springer.com/series/10032

Marcelo Reguero · Francisco Goin
Carolina Acosta Hospitaleche · Tania Dutra
Sergio Marenssi

Late Cretaceous/ Paleogene West Antarctica Terrestrial Biota and Its Intercontinental Affinities

 Springer

Marcelo Reguero
División Paleontología de Vertebrados
Museo de La Plata
La Plata
Argentina

Francisco Goin
División Paleontología de Vertebrados
Museo de La Plata
La Plata
Argentina

Carolina Acosta Hospitaleche
División Paleontología de Vertebrados
Museo de La Plata
La Plata
Argentina

Tania Dutra
SINOS-UNISINOS
Universidade do Vale do Rio dos
Sao Leopoldo
Rio Grande do Sul
Brazil

Sergio Marenssi
Instituto Antártico Argentino
Buenos Aires
Argentina

ISSN 2191-589X   ISSN 2191-5903   (electronic)
ISBN 978-94-007-5490-4   ISBN 978-94-007-5491-1   (eBook)
DOI 10.1007/978-94-007-5491-1
Springer Dordrecht Heidelberg New York London

Library of Congress Control Number: 2012945740

Printed on acid-free paper

Springer is part of Springer Science+Business Media (www.springer.com)

Preface

During the past two decades, geological explorations of the James Ross Basin, Weddell Sea, have revealed that this basin, located off the northeast tip of the Antarctic Peninsula (West Antarctica), contains one of the most important records of Late Cretaceous and Early Paleogene life in the Southern Hemisphere. The early explorer and scientist Otto Nordenskjöld (1905, p. 252), leader of the Swedish South Polar Expedition, envisioned the paleontologic and biogeographic importance of this basin "... *where, maybe, many animals and plants were first developed that afterwards found their way as far as to northern lands*". These discoveries have not only provided new insights into the geologic history of Antarctica, but they have also provided answers to questions about life in Southern Hemisphere that have puzzled naturalists since Charles Darwin's voyage on HMS *Beagle*.

The sedimentary sequence exposed in the James Ross Basin comprises a thick section of Coniacian, Turonian, Campanian, Maastrichtian, Paleocene, Eocene, and probably earliest Oligocene. They form the only marine sequence of this age interval that crops out in Antarctica. The high-latitude biota contained in both the Cretaceous and Paleogene beds is unusually rich and diverse and rivaled only by that from New Zealand and southeastern Australia.

After the Swedish South Polar Expedition (1901–1903), more than 40 years passed before the basin was scientifically visited again, this time by members of the Falkland Islands Dependencies Survey (now the British Antarctic Survey). The establishment of the Argentine station Marambio on Seymour Island in 1969 initiated the modern phase of geologic and paleontologic studies in the James Ross Basin.

The view that Antarctica and South America were connected by a long causeway between the West Antarctica (WANT) and southern South America in the Late Cretaceous through the late Paleocene, and that terrestrial vertebrates were able to colonize new frontiers using this physiographical feature, is almost certainly correct. One of the most intriguing palaeobiogeographical phenomena involving the last phase of the breakup of the supercontinent Gondwana concerns the close similarities and, in most cases, inferred sister-group relationships of a number of terrestrial vertebrate taxa recovered from uppermost Cretaceous and Paleogene deposits of Antarctic Peninsula (West Antarctica) and southern South

America (Magellanic Region and Patagonia). Dispersion of extinct vertebrates between South America and Australia across Late Cretaceous land bridges involving WANT predicts their presence during that interval on the latter. However, although knowledge of Late Cretaceous vertebrates from West Antarctica (and the Antarctic Peninsula) is still severely limited, it has expanded substantially over the last two decades and includes discoveries of avian and non-avian dinosaurs. Plants and small- to medium-sized, obligate terrestrial mammals (e.g., marsupials and meridiungulates) gained broad distribution across West Antarctica land mass prior to fragmentation and were isolated on the Antarctic Peninsula before the end of the Paleocene.

From an evolutionary perspective the close relationships between terrestrial taxa found in the James Ross Basin are difficult to accommodate because, by the beginning of the Late Cretaceous (99.6 Ma), the major Gondwanan continental blocks (South America, Africa, Antarctica, Mad agascar, the Indian subcontinent, and New Zealand–Australia) were well into the process of breakup and dispersion.

This paper deals with the hypothesis that the Antarctic Peninsula linked South America and West Antarctica in the Late Cretaceous until the beginning of the Paleogene and we review here through the plant and vertebrate fossil taxa the paleogeographical evolution of the southern continents and oceans for both earlier and later time intervals. This communication aims to explain the geological/geophysical problems inherent in the hypothesis and to explore the consequences for the biogeographical distributions of marine and terrestrial vertebrate faunas in West Antarctica.

Contents

1 Introduction 1
References 5

2 West Antarctica: Tectonics and Paleogeography 9
2.1 East Antarctica/West Antarctica and Gondwanan
Paleobiogeography 14
References 15

3 Late Cretaceous/Paleogene Stratigraphy in the James Ross Basin ... 19
3.1 Late Cretaceous 19
3.2 Paleogene 21
3.2.1 Cross Valley Formation 21
3.2.2 La Meseta Formation 22
References 23

**4 South America/West Antarctica: Pacific Affinities of the
Weddellian Marine/Coastal Vertebrates** 27
4.1 Late Cretaceous/Paleogene Marine Fossil Vertebrates
of the James Ross Basin 29
4.1.1 Neoselachii and Teleostei Fossil Fishes 29
4.1.2 Marine Reptiles 34
4.1.3 Whales 35
4.2 Weddellian Sphenisciformes: Systematics, Stratigraphy,
Biogeography and Phylogeny 36
References 49

5 The Terrestrial Biotic Dimension of West Antarctica 55
5.1 West Antarctica Paleoflora 55
5.2 Late Cretaceous Terrestrial Vertebrates of the James Ross Basin ... 65
5.2.1 Non Avian Dinosaurs 67
5.2.2 Avian Dinosaurs 72

5.3   Paleogene Terrestrial Vertebrates of the James Ross Basin   74
      5.3.1   Gondwanatheres   77
      5.3.2   Metatherians   77
      5.3.3   Xenarthrans   84
      5.3.4   "Insectivora"   85
      5.3.5   Litopterns   86
      5.3.6   Astrapotheres   88
5.4   Paleogene Environmental Reconstruction of the Cucullaea I
      (Ypresian) and Submeseta (Priabonian) Biotas   89
5.5   Correlation of the Cucullaea I Terrestrial Fauna with Early
      Paleogene Patagonian Faunas.   89
5.6   West Antarctic Terrestrial Biota and Its Intercontinental
      Relationships   97
References   99

Acknowledgments   111

Appendix   113

Chapter 1
Introduction

Abbreviations

ACC	Antarctic Circumpolar Current
EOB	Eocene–Oligocene boundary
IB/P/B	Institute of Biology, University of Białystok, Poland
Ma	Megannum
MLP	Vertebrate Paleontology collections, Museo de la Plata (UNLP), Argentina
Nd	Neodynium
SALMA	South American Land-Mammal Age
SAM	South America
SANU	South American native ungulates
SEM	Scanning electron micrograph
St	Strontium
WANT	West Antarctica

One of the most intriguing paleobiogeographical phenomena related to the final stage of Gondwanan breakup is the close similarities and, in most cases, inferred sister-group relationships, of a number of terrestrial and marine/coastal vertebrate taxa recovered from Paleogene deposits of West Antarctica with those from other continents (South America, Australia). These continents are today separated by large and deep ocean floors, which was not the case in the geological past. However, the inferred timing of continental separation does not always match with the inferred time of vertebrate dispersals.

The disjunct distributions of many groups of plants and animals in the Southern Hemisphere have largely been explained by vicariance. That the ancient floras of South America, New Zealand, Antarctica and Australia were shared is supported by fossil evidence (Raven and Axelrod 1972). However, recent studies, especially those based on molecular phylogenetics, have suggested that in many cases post-Gondwanan dispersal must have played an important role in determination of present day distributions of animals and plants. In some cases the distribution of groups of plants and animals demand land connections ("land bridges")

M. Reguero et al., *Late Cretaceous/Paleogene West Antarctica Terrestrial Biota and Its Intercontinental Affinities*, SpringerBriefs in Earth System Sciences, DOI: 10.1007/978-94-007-5491-1_1, © The Author(s) 2013

where none exists today. These land bridges may have been critical to intercontinental dispersal of land vertebrates. Also, eustatic changes in sea level can determine whether or not any areas of positive relief above the ocean floor became a
viable route, and for which interval of time. In the case of South America and West
Antarctica, the geologic histories of both were closely tied to subduction and clos-
ing of marginal basins along the southern margin of the Pacific Ocean.

Paleoclimates in Antarctica have greatly changed from the first (warmer) half
of the Cenozoic to the second (cooler) half. Even though its paleolatitude has
remained stationary since the Cretaceous, Antarctica became permanently glaciated only after the Eocene; a cool but not glacial early Cenozoic Antarctic climate
is well known (see, for example, Dingle and Lavelle 1998; Stilwell and Feldman
2000; Dutton et al. 2002). Evidence from an ever-increasing range of sources
(ice-rafted debris, Antarctic continental shelf drilling, marine benthic oxygen iso-
topes, clay mineralogy, deep-sea biotic changes and hiatuses indicating Southern-
Origin Bottom Water SOBW onset) point to the initiation of West Antarctic
glaciation at sea level close to the Eocene–Oligocene boundary (~34 Ma). Before
that, the northern tip of the Antarctic Peninsula and southernmost South America
(Patagonia, Argentina and Magallanic Region, Chile) were physically connected
allowing the dispersal of plants and animals between both areas (Olivero et al.
1991; Marenssi et al. 1994; Shen 1995; Reguero et al. 1998; Reguero et al. 2002;
Reguero and Marenssi 2010; Sallaberry et al. 2010; Bowman et al. 2012) (Fig. 1.1).

Paleobiogeographically it is useful to compare the wider patterns of biogeography
in the Southern Hemisphere (South America and Australia) to identify similarities
with the history of the Antarctic biota. This is because two of the important evolutionary forces within Antarctica, vicariance and dispersal, have often been treated
as competing explanations for the distribution of the Southern Hemisphere biota
(Sanmartín and Ronquist 2004). Populations of related terrestrial or freshwater
taxa that are now separated by oceans can be explained by either oceanic dispersal
or fragmentation of a previously contiguous land mass.

In order to understand the historical biogeography of West Antarctica, a thor-
ough understanding of the origin and gradual sundering of Gondwana landmasses
is required. Presently, the Antarctic Peninsula, as part of West Antarctica, is sur-
rounded by vast oceanic barriers on all sides but the South, where a thick ice-sheet
hides its relationship with the rest of this continent. It is closest to continental
Patagonia (South America), approximately 1000 km to the north, but lies 7500 km
from India, 7000 km from Madagascar and 6800 km from Australia. Since the
early Cretaceous, the Antarctic Peninsula has been located at almost the same
paleolatitude (South 60–65°; Lawver et al. 1992).

The Larsen Basin is located on the continental shelf off the coast of the
northern Antarctic Peninsula (Macdonald et al. 1988). The better-known James
Ross Basin (del Valle et al. 1992) is the northern sub-basin of the Larsen Basin
(Fig. 1.2). A 6–7 km-thick sedimentary succession related to the evolution of the
Larsen Basin, from a continental-rift to a back-arc setting, was deposited between
Jurassic and Eocene times in marine and transitional environments (Hathway
2000). The James Ross sequence is made up of marine clastic rocks deposited
within a back-arc basin (Elliot 1988; Hathway 2000) and represents the topmost

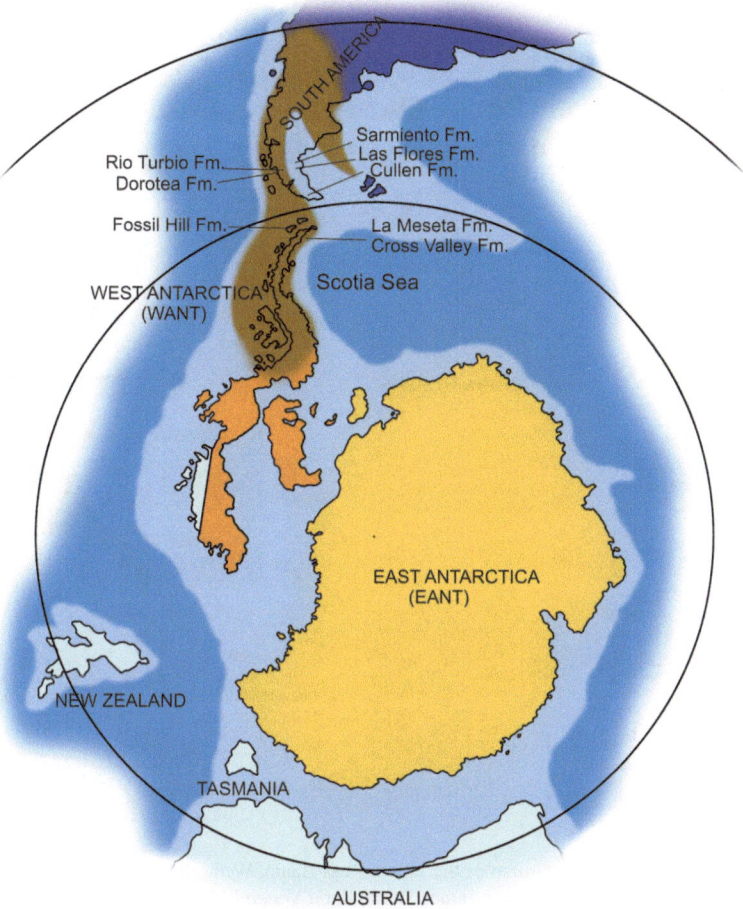

Fig. 1.1 Paleogene paleogeographic reconstruction of southern continents showing the location of the Paleocene and Eocene units discussed in the text. Compiled from distributional data after Zinsmeister (1982); Woodburne and Zinsmeister (1984); Lawver et al. (1992); Reguero et al. (2002). *Abbreviations* NZ, New Zealand

beds of a regressive mega-sequence (Pirrie et al. 1991). It includes an extremely thick sequence of Lower Cretaceous-Paleogene marine sedimentary rocks, which are divided into three main lithostratigraphic groups: the basal Gustav Group (Aptian-Coniacian), the intermediate Marambio Group (Santonian-Danian), and the upper Seymour Island Group (Paleogene) (e.g., Rinaldi 1982; Crame et al. 1991; Riding and Crame 2002) (Fig. 1.2).

Evidence of Paleogene Antarctic marine and terrestrial faunas come almost exclusively from the James Ross Basin (Reguero and Gasparini 2007) and it is almost exclusively based on fossils from the Eocene-?earliest Oligocene La Meseta Formation on Seymour (=Marambio) Island, situated east of the Antarctic Peninsula (Fig. 1.2), and secondly from the Eocene of the Fildes Peninsula, King

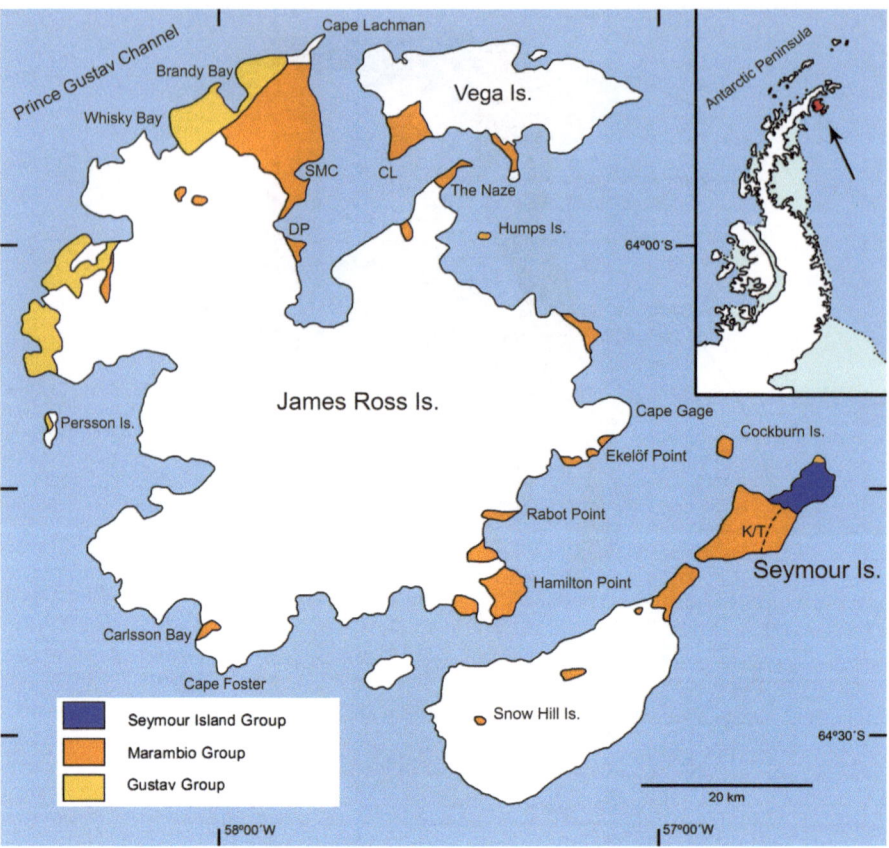

Fig. 1.2 Schematic geological map of the James Ross Basin, Weddell Sea, north-eastern Antarctic Peninsula. White areas are either the James Ross Island Volcanic Group or snow/ice cover. Locality key: DP, Dreadnought Point; BB, Brandy Bay; CL, Cape Lamb; SMC, Santa Marta Cove. Position of the Cretaceous-Paleogene boundary on Seymour Island indicated by the symbol K/T

George (=25 de Mayo) Island (Covacevich and Rich 1982; Jianjun and Shuonan 1994), and from Eocene glacial erratics from McMurdo Sound, East Antarctica (Stilwell and Zinsmeister 2000). The discovery of terrestrial fossil mammals in Antarctica has been recorded from the NW portion of Seymour Island, Antarctic Peninsula (Fig. 1.3). The terrestrial mammal-bearing unit, La Meseta Formation, represents sedimentation in coastal and shallow-marine environments. Although La Meseta Formation has generally been regarded as spanning a time period from early Eocene to ?Earliest Oligocene (Ivany et al. 2006), the fossiliferous rocks are exclusively Eocene in age (Montes et al. 2010).

In the last two decades, the James Ross Basin has been the scene of concentrated paleontological and geological fieldwork. Our knowledge of Paleogene vertebrate fauna has increased, but data are still sparse for the late Cretaceous and Paleocene history of vertebrates in Antarctica. In contrast, major progress

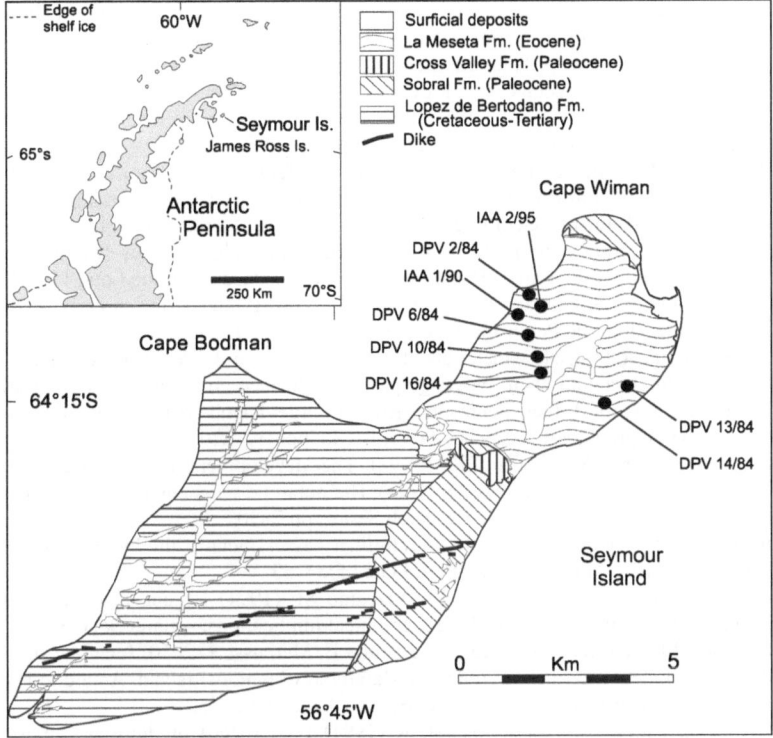

Fig. 1.3 Geological map of Seymour (Marambio) Island, Antarctic Peninsula, showing the IAA and DPV vertebrate-bearing localities cited in the text

and results have been obtained for the Eocene. This work summarizes information documenting major changes in the marine, coastal and terrestrial biota in the Eocene of the James Ross Basin before significant permanent ice sheets first appeared at the Eocene-Oligocene boundary in Antarctica. We present faunal data supporting the existence of a high latitude and altitude land biota showing some differences to related, penecontemporaneous faunas from Patagonia.

References

Bowman VC, Francis JE, Riding JB, Hunter SJ, Haywood AM (2012) A latest Cretaceous to earliest Paleogene dinoflagellate cyst zonation from Antarctica, and implications for phytoprovincialism in the high southern latitudes. Rev Palaeobot Palynol 171:40–56

Covacevich V, Rich PV (1982) New birds ichnites from Fildes Peninsula, King George Island, West Antarctica. In: Craddock C (ed) Antarctic Geoscience, University of Wisconsin Press. Madison, pp 245–254

Crame JA, Pirrie D, Riding JB, Thomson MRA (1991) Campanian-Maastrichtian (Cretaceous) stratigraphy of the James Ross island area, Antarctica. J Geol Soc Lond 148:1125–1140. doi:10.1144/gsjgs.148.6.1125 del Valle RA, Elliot DH, Macdonald DIM (1992) Sedimentary basins on the east flank of the Antarctic Peninsula: proposed nomenclature. Antarct Sci 4:477–478

Dingle R, Lavelle M (1998) Antarctic Peninsula cryosphere: early Oligocene (c. 30 Ma) initiation and a revised glacial chronology. J Geol Soc Lond 155:433–437

Dutton AL, Lohmann K, Zinsmeister WJ (2002) Stable isotope and minor element proxies for Eocene climate of Seymour Island Antarctica. Paleoceanography 17(2):1–13

Elliot DH (1988) Tectonic setting and evolution of the James Ross Basin, northern Antarctic Peninsula. In: Feldmann RM, Woodburne MO (eds) Geology and Paleontology of Seymour Island, Antarctic Peninsula. Geological Society of America, Memoir 169:541–555

Hathway B (2000) Continental rift to back-arc basin: Jurassic–Cretaceous stratigraphical and structural evolution of the Larsen Basin, Antarctic Peninsula. J Geol Soc Lond 157:417–432

Ivany LC, Van Simaeys S, Domack EW, Samson SD (2006) Evidence for an earliest Oligocene ice sheet on the Antarctic Peninsula. Geology 34(5):377–380

Jianjun L, Shuonan Z (1994) New materials of bird ichnites from Fildes Peninsula, King George Island of Antarctica and their biogeographic significance. In: Shen Y (ed) Stratigraphy and palaeontology of Fildes Peninsula, King George Island, Antarctica. State Antarctic Commitee, Monograph, vol 3. Science Press, Beijin, pp 239–249

Lawver LA, Gahagan LM, Coffin FM (1992) The development of palaeoseaway around Antarctica. In: Kennett JP, Warnke DA (eds) The antarctic paleoenvironment: a perspective on global change, vol 65. Antarctic Research Series, pp 7–30

MacDonald DIM, Baker PF et al (1988) A preliminary assessment of the hydrocarbon potential of the Larsen Basin, Antarctica. Mar Petrol Geol 5:34–53

Marenssi SA, Reguero MA, Santillana SN, Vizcaíno SF (1994) Eocene land mammals from Seymour Island, Antarctica: palaeobiogeographical implications. Antarct Sci 6:3–15

Montes M, Nozal F, Santillana S, Tortosa F, Beamud E, Marenssi S (2010) Integrate stratigraphy of the Upper Paleocene-Eocene strata ofs Marambio (Seymour) Island, Antarctic Peninsula. XXXI SCAR, Open Science Conference, Buenos Aires, Argentina

Olivero E, Gasparini Z, Rinaldi C, Scasso R (1991) First record of dinosaurs in Antarctica (Upper Cretaceous, James Ross Island): paleogeographical implications. In: Thomson MRA, Crame JA, Thomson JW (eds) Geological Evolution of Antarctica. Cambridge University Press, Cambridge, pp 617–622

Pirrie D, Crame JA, Riding JB (1991) Late Cretaceous stratigraphy and sedimentology of Cape Lamb, Vega Island, Antarctica. Cretaceous Research 12:227–258

Raven PH, Axelrod DI (1972) Plate tectonics and Australasian paleobiogeography. Science 176:1379–1386

Reguero MA, Gasparini Z (2007) Late Cretaceous-Early Tertiary marine and terrestrial vertebrates from James Ross Basin, Antarctic Peninsula: a review. In: Rabassa J, Borla ML (eds). Antarctic Peninsula and Tierra del Fuego: 100 years of Swedish-Argentine scientific cooperation at the end of the world. Taylor Francis, London, pp 55–76

Reguero MA, Marenssi SA (2010) Paleogene climatic and biotic events in the terrestrial record of the Antarctic Peninsula: an overview. In: Madden R, Carlini AA, Vucetich MG, Kay R (eds) The paleontology of Gran Barranca: evolution and environmental change through the Middle Cenozoic of Patagonia. Cambridge University Press, Cambridge, pp 383–397

Reguero MA, Vizcaíno SF, Goin FJ, Marenssi SA, Santillana SN (1998) Eocene high-latitude terrestrial vertebrates from Antarctica as biogeographic evidence. In: Casadio S (ed) Paleógeno de América del Sur y de la Península Antártica. Asociación Paleontológica Argentina, Publicación Especial, vol 5. pp 185–198

Reguero MA, Marenssi SA, Santillana SN (2002) Antarctic Peninsula and Patagonia Paleogene terrestrial environments: biotic and biogeographic relationships. Palaeogeogr Palaeoclimatol Palaeoecol 179:189–210

Riding JB, Crame JA (2002) Aptian to Coniacian (early-late Cretaceous) palynostratigraphy of the Gustav Group, James Ross Basin, Antarctica. Cretac Res 23:739–760

Rinaldi CA (1982) The upper cretaceous in the James Ross Island group. In: Craddock C (ed) Antarctic geoscience. The University of Wisconsin Press, Madison, pp 331–337

Sallaberry MA, Yury-Yáñez RE, Otero RA, Soto-Acuña S, Torres GT (2010) Eocene birds from the western margin of southernmost South America. J Paleontol 84(6):1061–1070

Sanmartín L, Ronquist F (2004) Southern hemisphere biogeography inferred by event-based models: plant versus animal patterns. Syst Biol 53(2):216–243

Shen Y (1995) Subdivision and correlation of Cretaceous to Paleogene volcano-sedimentary sequence from Fildes Peninsula, King George Island, Antarctica. In: Shen Y (ed) Stratigraphy and Palaeontology of Fildes Peninsula, King George Island, Antarctica. State Antarctic Committee, Monograph. vol 3. Science Press, Beijin, pp 1–36

Stilwell JD, Feldman RM (2000) Paleobiology and Paleoenvironments of Eocene Rocks, McMurdo Sound. East Antarctica. Ant. Res. Ser American Geophysical Union, Washington, DC

Stilwell JD, Zinsmeister WJ (2000) Paleobiogeographic synthesis of the Eocene macrofauna from McMurdo Sound, Antarctica. In: Stilwell JD, Feldmann RM (eds) Paleobiology and paleoenvironments of Eocene rocks: McMurdo Sound, East Antarctica. Antarct Res Ser 76:365–372, AGU, Washington, doi:10.1029/AR076p0365

Woodburne MO, Zinsmeister WJ (1984) The first land mammal from Antarctica and its biogeographic implications. J Paleontol 54:913–948

Zinsmeister WJ (1982) Late Cretaceous-Early Tertiary molluscan biogeography of southern Circum-Pacific. J Paleontol 56:84–102

Chapter 2
West Antarctica: Tectonics and Paleogeography

The origin of West Antarctica (WANT) can be traced back to the Terra Australis orogenesis that began between 520 Ma and 510 Ma—shortly after the terminal suturing of Gondwana (Boger 2011). The onset of this event was responsible for the termination of passive margin sedimentation along much of the Pacific margin of Gondwana and marks the beginning of widespread and broadly coeval deformation and arc-type plutonism. It also began a long-lived process of accretion that added much of the crust that defines eastern Australia, West Antarctica (domain 5 of Boger 2011, Fig. 2.1), and western South America (Cawood 2005, 2009).

Post-Gondwana accretionary growth—the Terra Australis and Gondwanide Orogenies—The suturing of the West Gondwana and Australo–Antarctic plates along the Kuunga Orogen brought to an end the long-lived process of convergence between the pre-collision components of Gondwana. The result was a reconfiguration of the early to middle Cambrian plate system and the consequent transfer of ocean floor consumption from between the pre-Gondwana cratons to the outboard Pacific margin of newly formed Gondwana supercontinent (Cawood 2005). This led to the establishment of the accretionary Terra Austrais Orogen (Cawood 2005), a general name given to the orogenic belt that stretched continuously from northern South America to northern Australia and which began in the early to middle Cambrian and lasted until the late Carboniferous. In Antarctica Terra Australis (Ross) orogenesis also deformed and variably metamorphosed the pre-Gondwana passive margin (Fig. 2.2).

The Gondwana supercontinent underwent a sequential fragmentation over approximately 165 Ma. India and Madagascar may have maintained a link with Antarctica via Sri Lanka until the end of the Aptian (112 Ma) but by the earliest Albian a deep seaway separated India/Madagascar/Sri Lanka from East Antarctica. The continental separation between West Antarctica and Campbell Plateau (the first appearance of ocean crust) was produced at around 83 Ma (Larter et al. 2002). Rifting was thus achieved via a two-stage process—extension and crustal thinning between Australia–East Antarctica and New Zealand–West Antarctica between approximately 105 Ma and 95 Ma, followed by extension and then

M. Reguero et al., *Late Cretaceous/Paleogene West Antarctica Terrestrial Biota and Its Intercontinental Affinities*, SpringerBriefs in Earth System Sciences, DOI: 10.1007/978-94-007-5491-1_2, © The Author(s) 2013

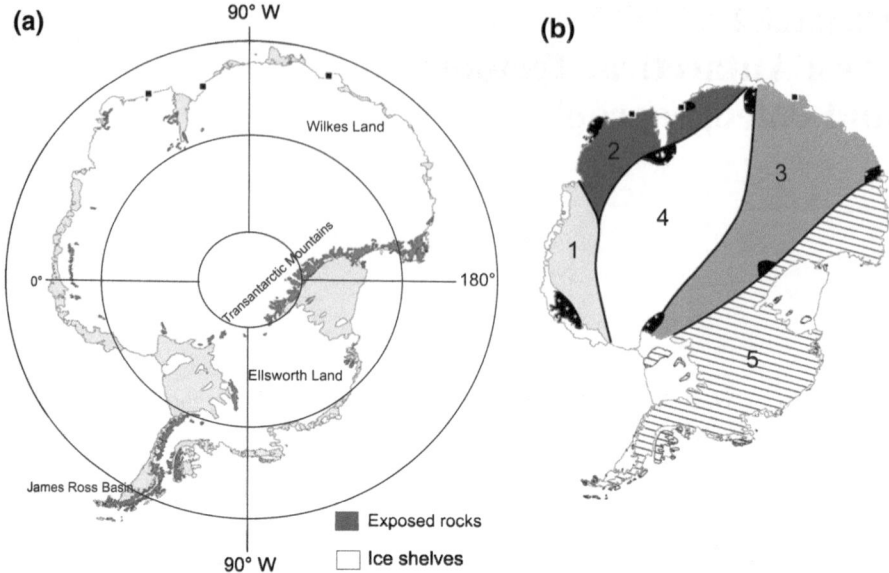

Fig. 2.1 Present day Antarctica: **a** Distribution of rock exposures and major geographic place names; **b** Tectonic domains of Antarctica differentiated on the basis of their affinities with Antarctica's correlatives within Gondwana (after Boger 2011)

separation along a slightly oblique axis between West Antarctica–Australia and New Zealand (plus related blocks) by 83 Ma (Kula et al. 2007). The separation of New Zealand from the margin of West Antarctica effectively formed the modern Antarctic continent as we recognize it today. Since rifting, the Antarctic plate has remained effectively stationary and geological activity has been limited to ongoing extension and volcanism within the West Antarctic rift system. Most deformation occurred between 43 Ma and 26 Ma during which time another 180 km of extension occurred (Cande et al. 2000). Deformation at this time is commonly linked to uplift and 4–9 km of denudation along the rift flank now defined by the Transantarctic Mountains (e.g. Fitzgerald and Gleadow 1988; Fitzgerald and Stump 1997; Miller et al. 2010).

West Antarctica is complex in two ways: first, its margin (together with that of Zealandia) has undergone convergence and divergence; second, West Antarctica itself is a complex of microcontinental blocks that have moved independently of each other and have been widely extended over the past 200 million years in the West Antarctic rift system. West Antarctica is thus an ensemble of blocks (shaded) that have moved independently of each other and of cratonic East Antarctica. The blocks of West Antarctica (Dalziel and Lawver 2001, Fig. 2.2) are crossed by the West Antarctic Rift System between the Bellingshausen Sea and the Ross Sea. The rift trough between the rift faults bounding the Transantarctic Mountains and the Ellsworth–Whitmore Mountains block on one side and Marie Byrd Land on the other contains basins in the Ross Sea (LeMasurier 2008). The volcanic islands around Antarctica are described by Quilty (2007).

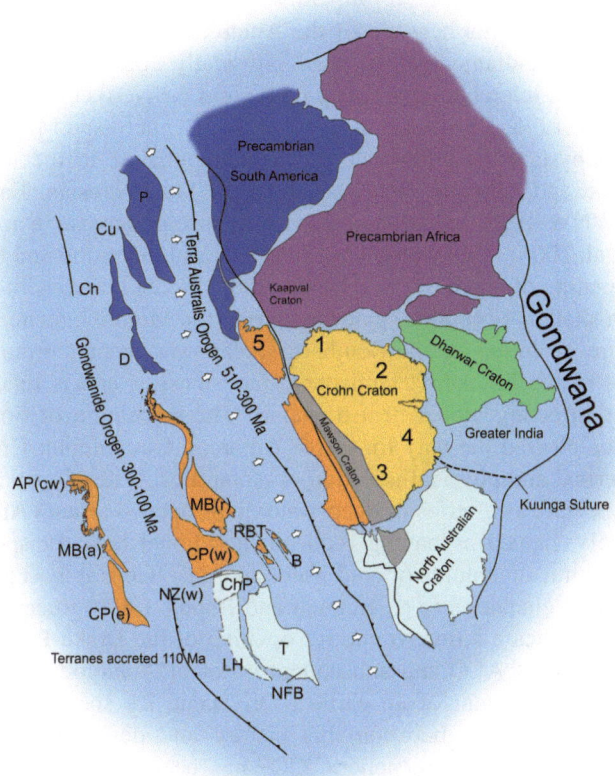

Fig. 2.2 Antarctic palaeogeography from the middle Cambrian to the late Carboniferous. Westward subduction under the Pacific margin of Gondwana marks onset of Terra Australis orogenesis. Black stars mark exposures of the passive margin sequences: EM = Ellsworth–Whitmore Mountains, Ch = Chilenia Terrane, ChP = Challenger Plateau (western domain New Zealand (NZ(w))), CP(w) = Campbell Plateau (western domain New Zealand), Cu = Cuyania Terrane, D = Deseado Terrane, LH = Lord Howe Rise, MB(r) = Ross Province Marie Byrd Land, NEFB = New England Fold Belt, P = Pampean Terrane, RBT = Robertson Bay Terrane and T = Tasmanides. Outboard terranes: AP(cw) = central and western domains of the Antarctic Peninsula, CP (e) = Campbell Plateau (eastern domain New Zealand) and MB(a) = Amundsen Province Marie Byrd Land. Numbered regions (1–5) refer to the tectonic domains shown in Fig. 2.1

By the Late Cretaceous, the Antarctic Peninsula and the remainder of West Antarctica are believed to have been comprised of a number of discrete microcontinental blocks forming a single elongated landmass that extended southward from southern South America. Since at least that time the Antarctic Peninsula has been in its present position relative to South America, at almost the same paleolatitude (South 60–65°) (Lawver et al. 1992), but became glaciated only more recently: a cool but not glacial early Cenozoic Antarctic climate is well known (see, for example, Dingle et al. 1998; Stilwell and Feldman 2000; Dutton et al. 2002). Before that, the northern tip of the Antarctic Peninsula and southernmost South America (Magallanic region) were physically connected facilitating both florirstic and faunal interchange between both areas (Olivero et al. 1991; Marenssi et al. 1994; Shen 1995; Reguero et al. 1998, 2002; Reguero and Marenssi 2010).

Marine geophysical data demonstrates a major change in the motion of the South American and Antarctic plates at about 50 Ma (Ypresian, Early Eocene), from N–S to WNW–ESE, accompanied by an eightfold increase in separation rate (Livermore et al. 2005). This would have led to crustal extension and thinning, and perhaps the opening of small oceanic basins of the Weddell Sea, with the probable formation of a shallow (<1000 m) gateway during the Middle Eocene.

The climatic evolution of the Southern Ocean was in part brought about by rearrangement of the Southern Hemisphere land masses, and is related to circulation changes affecting global heat transport. The separation of South America from Antarctica cleared the way for a free Antarctic Circumpolar Current (ACC) to produce thermal and physical isolation of Antarctica. It developed at some time during the Cenozoic, as a series of deep-water gaps opened around Antarctica, and has been widely viewed as having reduced meridional heat transport, isolating the continent within an annulus of cold water and thus being at least partly responsible for Antarctic glaciation (e.g., Kennett and Barker 1990). Recent articles provide data which indicate a time frame for opening of the Drake Passage resulting in the development of ACC and subsequent onset of Antarctic climatic cooling, leading to the development of an earliest Oligocene ice sheet on the Antarctic Peninsula. The correlation between this history and determinations of falling ancient partial pressure of atmospheric CO_2 has been linked to global cooling (Pearson and Palmer 2000; Pagani et al. 2005). Livermore et al. (2005) propose a shallow water opening (<1,000 m deep) as early as 50 Ma based on tectonic evidence in the Weddell Sea. They correlated the formation of new tectonic basins in the area that will later become the Scotia Arc and the northern Antarctic Peninsula, with a drop in southern ocean temperatures based on oxygen isotope data from benthic foraminifera to represent the initial shallow water opening between South America and Antarctica. Eagles et al. (2006) indicate that subsidence in the area of some small oceanic basins in the southern Scotia Sea (Dove Basin and Protector Basin) east of Drake Passage, was underway at ~50 Ma, producing a deepening shallow- or intermediate- depth rift that gave way to seafloor spreading in the Dove basin by around 41 Ma. Basin opening coincided with a sustained period of global cooling that started after the Ypresian (Early Eocene). Extension in the region of the Dove and Protector basins would have opened Drake Passage to shallow or intermediate depth oceanic circulation between the Pacific and Atlantic oceans for the first time by 41 Ma (Fig. 2.3).

Scher and Martin (2006) utilized a rare earth element neodymium (Nd) contained in fish teeth extracted from deep sea cores taken in the south Atlantic Ocean between southern Africa and Queen Maud Land on the Antarctic craton to provide data on a deepwater opening of the Drake Passage. They used ratios of $^{143}Nd/^{144}Nd$ to determine the transition from non-radiogenic to radiogenic

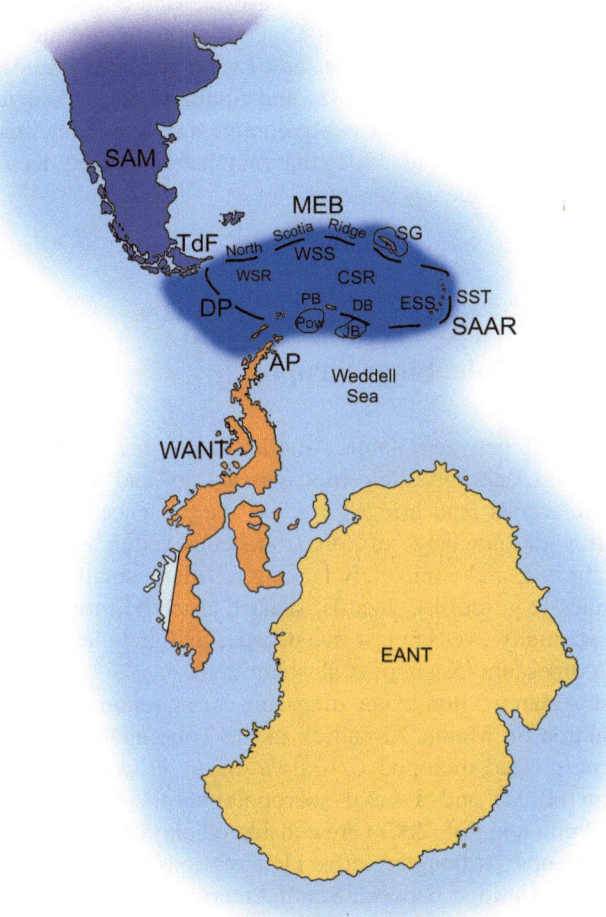

Fig. 2.3 Scotia and Weddell seas (Smith and Sandwell 1997). **AP**, Antarctic Peninsula; **CSS**, central Scotia Sea; **DP**, Drake Passage; **ESS**, East Scotia Sea; **MEB**, Maurice Ewing Bank; **PB**, Protector Basin; **Pow**, Powell Basin; **JB**, Jane Basin; **SAAR**, South American–Antarctic Ridge; **SG**, South Georgia; **SST**, South Sandwich Trench; **TdF**, Tierra del Fuego; **WSR**, West Scotia Ridge; **WSS**, west Scotia Sea

Nd values which would mark the flow of the more radiogenic Pacific waters into the Atlantic Ocean as the signal for the deepwater opening of the Drake Passage in the late Eocene (ca. 41 Ma). Variations of Nd (Neodynium) isotope ratios (^{143}Nd/^{144}Nd) in the Atlantic sector outline the history of the opening of Drake Passage more directly than previous studies that relied on other paleoceanographic proxies to infer a Pacific–Atlantic connection. These variations suggest an influx of shallow Pacific seawater approximately at 41 Ma. Subsequent eNd increases during the late Eocene most likely represent progressive widening and deepening of the gateway. A pronounced eNd increase at Maud Rise ~37 Ma has been previously interpreted as an opening of Drake Passage to intermediate depths. For the Drake Passage region, there are at least two continental fragments, the South Orkney block (King and Barker 1988) and South Georgia (Dalziel et al. 1975; Ramos 1996), as well as other possibly continental, high-standing blocks in the central Scotia Sea (Barker et al. 1991) that may have impeded deep water circulation through Drake Passage.

2.1 East Antarctica/West Antarctica and Gondwanan Paleobiogeography

Vertebrate dispersal between South America and other Gondwanan continents or subcontinents (Australia, Indo-Madagascar) to and from West Antarctica across Late Cretaceous land bridges predicts the presence of their fossils in that continent. Although knowledge of Late Cretaceous vertebrates from Antarctica is still severely limited, and completely lacking for many small terrestrial and freshwater vertebrates (e.g. tortoise, lizards, snakes, crocodyliforms, mammals), it has expanded substantially over the last two decades and includes discoveries of avian and non-avian dinosaurs (Reguero et al. submitted).

The oldest Antarctic non-avian dinosaurs were found in the Early Jurassic Hanson Formation on Mount Kirpatrick in the Transantarctic Mountains in East Antarctica. The crested theropod *Cryolophosaurus ellioti* described by Hammer and Hickerson (1994) and a basal sauropodomorph, *Glacialisaurus hammeri*, described by Smith and Pol (2007) are considered part of an early Gondwana radiation of these groups. Although relative plate motions suggest that there could not have been any land bridge between South America, Africa, Antarctica, India and Madagascar after 108–112 Ma, Sereno and Brusatte (2008) based on the phylogenetic relationship of Abelisauroidea to Carnotaurinae and the finding of a abelisaurid fossil from the Cenomanian of Niger, proposed the existence of land bridges between South America and Africa and between India and East Antarctica at 97 Ma. By the mid–Late Cretaceous (latter part of the Santonian, end-Cretaceous Normal Superchron, 83.5 Ma), South America, Antarctica and Australia were the only Gondwanan continents that were still connected; the Indian subcontinent had begun its separation from Madagascar c. 5 Ma earlier (e.g., Storey et al. 1995).

Plate tectonics and the terrestrial vertebrate fossil record from the James Ross Basin suggest dispersal of vertebrate groups from South America (Patagonia) to Antarctica and possibly onto Australia until as late as the Eocene. Woodburne and Case (1996) assumed that there may have been a terrestrial link between South America and Australia via Antarctica but it ended by 64 Ma when the South Tasman Rise submerged. Late Cretaceous terrestrial vertebrates from West Antarctica are thus of interest to address a variety of questions, ranging from the dispersal patterns of major vertebrate clades to Gondwanan endemism.

A new South Polar Province based on Maastrichtian to Danian dinoflagellate cysts proposed by Bowman et al. (2012) excludes most southern South American marine palynofloras. This interpretation supported by models of ocean currents around Antarctica implies an unrestricted oceanic connection across Antarctica between southern South America and the Tasman Sea. This hypothesis does not therefore support the presence of a continuous geographical landmass through South America, the Antarctic Peninsula and West Antarctica to Australasia (e.g., Woodburne and Zinsmeister 1984; Case et al. 1987; Case 1988; Cantrill and Poole 2002). However, this does not preclude the presence of closely spaced islands forming an archipelago through this region to account for the dispersion of *Nothofagus* spp. and marsupials across Antarctica from South America (Schuster 1976; Case 1988).

References

Barker PF, Dalziel IWD, Storey BC (1991) Tectonic development of the Scotia Arc region. In: Tingey (ed) Antarctic geology, Oxford University Press, Oxford, pp 215–248

Boger SD (2011) Antarctica—before and after Gondwana. Gondwana Res 19:335–371

Bowman VC, Francis JE, Riding JB, Hunter SJ, Haywood AM (2012) A latest Cretaceous to earliest Paleogene dinoflagellate cyst zonation from Antarctica, and implications for phytoprovincialism in the high southern latitudes. Rev Palaeobot Palynol 171:40–56

Case JA (1988) Paleogene floras from Seymour Island, Antarctic Peninsula. In: Feldmann RM, Woodburne MO (eds) Geology and Paleontology of Seymour Island, Antarctic Peninsula, vol 169. Memoir of the Geological Society of America, Boulder, pp 523–530

Case JA, Woodburne MO, Chaney D (1987) A gigantic phororhacoid (?) bird from Antarctica. J Paleontol 16:1280–1284

Cantrill DJ, Poole I (2002) Cretaceous patterns of floristic change in the Antarctic Peninsula. Geol Soc Lond Spec Publ 194:141–152

Cawood PA (2005) Terra Australis Orogen: Rodinia breakup and development of the Pacific and Iapetus margins of Gondwana during the Neoproterozoic and Paleozoic. Earth Sci Rev 69:249–279

Cawood PA, Kröner A, Collins WJ, Kusky TM, Mooney WD, Windley BF (2009) Accretionary orogens through earth history. In: Cawood PA, Kröner A (eds) Earth accretionary systems in space and time, vol 318. Geological Society, London, Special Publication, pp 1–36

Cande SC, Stock JM, Müller RD, Ishihara T (2000) Cenozoic motion between East and West Antarctica. Nature 404:145–150

Dalziel IWD, Dott RH Jr, Winn RD Jr, Bruhn RL (1975) Tectonic relations of South Georgia island to the southernmost Andes. Geol Soc Am Bull 86:1034–1040

Dalziel IWD, Lawver LA (2001) The lithospheric setting of the West Antarctic ice sheet. Am Geophys Union Antarct Res Ser 77:29–44

Dingle R, Marenssi S, Lavelle M (1998) High latitude Eocene climate deterioration: evidence from the northern Antarctic Peninsula. J S Am Earth Sci 11:571–579

Dutton AL, Lohmann K, Zinsmeister WJ (2002) Stable isotope and minor element proxies for Eocene climate of Seymour Island Antarctica. Paleoceanography 17(2):1–13

Eagles G, Livermore RA, Morris P (2006) Small basins in the Scotia Sea: the Eocene Drake passage gateway. Earth Planet Sci Lett 242:343–353

Fitzgerald PG, Gleadow AJW (1988) Fission-track geochronology, tectonics and structure of the Transantarctic Mountains in northern Victoria land, Antarctica. Chem Geol 73:169–198

Fitzgerald PG, Stump E (1997) Cretaceous and Cenozoic episodic denudation of the Transantarctic Mountains, Antarctica: new constraints from apatite fission track thermochronology in the Scott Glacier region. J Geophys Res 102:7747–7765

Hammer WR, Hickerson WJ (1994) A crested theropod dinosaur from Antarctica. Science 264:828–830

King EC, Barker PF (1988) The margins of the South Orkney Microcontinent. J Geol Soc Lond 145:317–331

Kula J, Tulloch A, Spell TL, Wells ML (2007) Two-stage rifting of Zealandia–Australia–Antarctica: evidence from $^{40}Ar/^{39}Ar$ thermochronometry of the sisters shear zone, Stewart Island, New Zealand. Geology 35:411–414

Kennett JP, Barker PF (1990) Latest Cretaceous to Cenozoic climate and oceanographic developments in the Weddell Sea, Antarctica: an ocean-drilling perspective. In: Barker PF, Kennett JP (eds) Proceedings of ODP Science Results, vol 113. pp 937–960

LeMasurier WE (2008) Neogene extension and basin deepening in the West Antarctic rift inferred from comparisons with the East African rift and other analogs. Geology 36:247–250

Lawver LA, Gahagan LM, Coffin FM (1992) The development of palaeoseaway around Antarctica. In: Kennett JP, Warnke DA (eds) The Antarctic paleoenvironment: a perspective on global change, vol 65. Antarctic Research Series, Antarctic Sea Ice, pp 7–30

Larter RD, Cunningham AP, Barker PF, Gohl K, Nitsche FO (2002) Tectonic evolution of the Pacific margin of Antarctica, 1. Late Cretaceous tectonic reconstructions. J Geophys Res 107:2345. doi:10.1029/2000JB000052

Livermore R, Nankivel A, Eagles G, Morris P (2005) Paleogene opening of Drake passage. Earth Planet Sci Lett 236:459–470

Miller SR, Fitzgerald PG, Baldwin SL (2010) Cenozoic range-front faulting and development of the Transantarctic Mountains near Cape Surprise, Antarctica: thermochronologic and geomorphologic constraints. Tectonics 29(TC1003): 21 doi:10.1029/2009TC002457

Marenssi SA, Reguero MA, Santillana SN, Vizcaíno SF (1994) Eocene land mammals from Seymour Island, Antarctica: palaeobiogeographical implications. Antarct Sci 6:3–15

Olivero E, Gasparini Z, Rinaldi C, Scasso R (1991) First record of dinosaurs in Antarctica (Upper Cretaceous, James Ross Island): paleogeographical implications. In: Thomson MRA, Crame JA, Thomson JW (eds) Geological Evolution of Antarctica. Cambridge University Press, Cambridge, pp 617–622

Pagani M, Zachos JC, Freeman KH, Tipple B, Bohaty S (2005) Marked decline in atmospheric carbon dioxide concentrations during the Paleogene. Science 309:600–603. doi:10.1126/science.1110063

Pearson PN, Palmer MR (2000) Atmospheric carbon dioxide concentrations over the past 60 million years. Nature 406:695–699

Quilty PG (2007) Origin and evolution of the sub-Antarctic islands: the foundation. Pap Proc R Soc Tasmania 141:35–58

Reguero MA, Vizcaíno SF, Goin FJ, Marenssi SA, Santillana SN (1998) Eocene high-latitude terrestrial vertebrates from Antarctica as biogeographic evidence. In: Casadio S (ed) Paleógeno de América del Sur y de la Península Antártica, vol 5. Asociación Paleontológica Argentina, Publicación Especial, pp 185–198

Reguero MA, Marenssi SA (2010) Paleogene climatic and biotic events in the terrestrial record of the Antarctic Peninsula: an overview. In: Madden R, Carlini AA, Vucetich MG, Kay R (eds) The paleontology of Gran Barranca: evolution and environmental change through the middle Cenozoic of Patagonia, Cambridge University Press, pp 383–397

Ramos VA (1996) Geología de las Islas Georgias del Sur. In: Ramos VA, Turic MA (eds) Geología y Recursos Naturales de la Plataforma Continental Argentina. XIII Congreso Geologico Argentino y III Congreso de Exploracion de Hidrocarburos, Buenos Aires, pp 359–368

Scher HD, Martin EE (2006) The timing and climatic influence of the opening of Drake passage. Science 312:428–430

Schuster RM (1976) Plate tectonics and its bearing on the geographical origin and dispersal of angiosperms. In: Beck CB (ed) Origin and early evolution of angiosperms. Columbia University Press, New York, pp 48–138

Stilwell JD, Feldman RM (2000) Paleobiology and paleoenvironments of Eocene rocks, McMurdo Sound East Antarctica. Ant Res Ser American Geophysical Union, Washington

Shen Y (1995) Subdivision and correlation of Cretaceous to Paleogene volcano-sedimentary sequence from Fildes Peninsula, King George Island, Antarctica. In: Shen Y (ed) Stratigraphy and palaeontology of Fildes Peninsula, King George island, Antarctica. State Antarctic Committee, Monograph, vol 3. Science Press, Beijin, pp 1–36

Smith ND, Pol D (2007) Anatomy of a basal sauropodomorph dinosaur from the early Jurassic Hanson formation of Antarctica. Acta Palaeontol Pol 52:657–674

Sereno PC, Brusatte SL (2008) Basal abelisaurid and carcharodontosaurid theropods from the Lower Cretaceous Elrhaz Formation of Niger. Acta Palaeontol Pol 53:15–46

Storey M, Mahoney JJ, Saunders AD, Duncan RA, Kelley SO, Coffin MF (1995) Timing of hotspot related volcanism and the breakup of Madagascar and India. Science 267:852–855

Woodburne MO, Case JA (1996) Dispersal, vicariance and the late Cretaceous to early tertiary land mammal biogeography from South America to Australia. J Mamm Evol 3:121–161

Woodburne MO, Zinsmeister WJ (1984) The first land mammal from Antarctica and its biogeographic implications. J Paleontol 54:913–948

Chapter 3
Late Cretaceous/Paleogene Stratigraphy in the James Ross Basin

3.1 Late Cretaceous

Late Cretaceous sedimentary rocks are only exposed around the northern part of West Antarctica, on the South Shetland Islands and James Ross Island Group, Weddell Sea (Fig. 3.1). They were deposited in very different tectonic settings and environments. The South Shetland Island sequence represents a Cenozoic outer-arc (Birkenmajer 1995) or fore-arc (Elliot 1988) succession composed mainly of terrestrial volcanic and sedimentary deposits. It should also be noted that some very thin and intermittent sequences of terrestrial sedimentary rocks interbedded within extensive volcanic units on King George Island, South Shetland Islands may be of Maastrichtian age. For example, at Half Three Point, Fildes Peninsula (62°13′40″S; 58°59′01″W), a 4-m-thick sequence of tuffs, tuffaceous sandstones, siltstones and mudstones has yielded an Rb–Sr age of 71.3 ± 0.3 Ma (Dutra and Batten 2000). On the other hand, the James Ross sequence is made up of marine clastic rocks deposited within a back-arc basin (Elliot 1988; Hathway 2000) and representing the topmost beds of a regressive mega-sequence (Pirrie et al. 1991).

The James Ross Basin includes an extremely thick sequence of Lower Cretaceous-Palaeogene marine sedimentary rocks, and is divided into three principal lithostratigraphic groups: the basal Gustav (Aptian-Coniacian), the intermediate Marambio (Santonian-Danian), and the upper Seymour Island (Paleogene) groups (e.g., Rinaldi 1982; Crame et al. 1991; Riding and Crame 2002).

The 2,100 m-thick marine sediments of the Gustav Group are exposed along the NW Coast of James Ross Island (Fig. 3.1). The coarse-grained lower three units of the group (Lagrelius Point, Kotick Point and Whisky Bay formations) were deposited in a deep marine setting and represent proximal submarine-fan and slope-apron depositional systems (Ineson 1989; Buatois and López-Angrinman 1992). The fourth and younger unit, Hidden Lake Formation, is composed of distinctive brown-weathering, volcaniclastic sediments corresponding to a fan delta that are readily traced along the length of NW James Ross Island (Whitham et al. 2006).

M. Reguero et al., *Late Cretaceous/Paleogene West Antarctica Terrestrial Biota and Its Intercontinental Affinities*, SpringerBriefs in Earth System Sciences, DOI: 10.1007/978-94-007-5491-1_3, © The Author(s) 2013

Fig. 3.1 Stratigraphic column of La Meseta Formation on Seymour Island, Antarctic Peninsula (modified from Montes et al. 2010). Strontium date values from Dingle Lavelle 1998; Dutton et al. 2002; Reguero et al. 2002 and Ivany 2008. *Abbreviations cu. Cucullaea, n. naticids, v. veneroids, and t. Turritella*

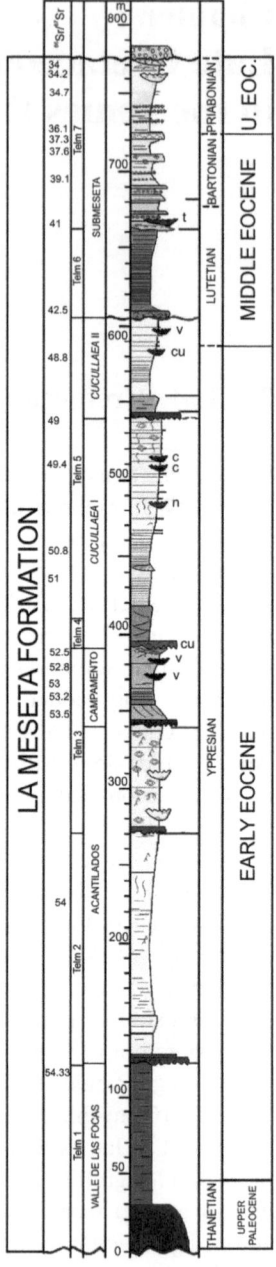

The finer-grained Marambio Group, about 3000 m thick, is exposed in northern James Ross, Vega, and Humps islands in the north as well as in southern James Ross, Snow Hill, Seymour and Cockburn islands (Olivero et al. 1986, 1992; Pirrie et al. 1991, 1997; Olivero 2012). Overall the Marambio Group represents the construction of a prograding shelf extending into the Weddell Sea. The expansion of the shelf area was punctuated by three major sedimentary cycles: the N (Santonian-early Campanian); NG (late Campanian-early Maastrichtian); and MG (early Maastrichtian-Danian) sequences (Olivero 2012).

3.2 Paleogene

The Paleogene back-arc deposits comprise more than 1,000 meters of shallow marine to coastal fossiliferous clastic sedimentary rocks mainly of Paleocene and Eocene ages (Elliot 1988; Sadler 1988; Marenssi et al. 1998a). They are exposed on Seymour and Cockburn islands approximately 100 km SE of the northern tip of the Antarctic Peninsula and represent the uppermost part of the James Ross Basin (del Valle et al. 1992) succession (Fig. 3.1).

The James Ross Basin (del Valle et al. 1992) on the northeast continental shelf of the Antarctic Peninsula consists of exposed? Barremian-? earliest Oligocene sedimentary rocks deposited in a retroarc setting. The Tertiary section exposed mainly on Seymour Island includes the Late Paleocene unit Klb 10 of the López de Bertodano Formation, Sobral Formation, Cross Valley Formation and the richly fossiliferous Eocene-?earliest Oligocene La Meseta Formation, all of them deposited in incised valley settings. The last two units, Cross Valley and La Meseta formations, yield Weddellian spheniSciforms and fishes.

3.2.1 Cross Valley Formation

At its type section in the middle part of Seymour Island, the Cross Valley Formation (Elliot and Trautman 1982) fills a steep-sided valley cut in the Lower Paleocene Sobral Formation and older beds. The width of the "canyon" is 600 m at a maximum and the valley fill consists of more than 120 m of coarse sands and axial mass flows deposits overlain by some 40 m thick post-valley fill finer-grained beds. The base of this unit is a concave-up unconformity oriented in a NW–SE direction with steep, asymmetrical sides. The southern edge dips 45° towards the North while the northern margin plunges 35° to the South. The top of the unit is another unconformity at the base of the La Meseta Formation. The younger beds cut by the basal unconformity are dated as Danian while the oldest sediment covering these units are regarded Early Eocene. Dinoflagellates and pollen collected from the upper part of the Cross Valley Formation also suggest a Late Paleocene age for this unit (Askin 1988; Wrenn and Hart 1988).

The Cross Valley Formation has been divided into three informal members (Marenssi et al. 2012). The lower Cerro Arañado member comprises coarse-grained pebbly volcaniclastic sandstones with carbonised wood (Tpcv1 and two of Sadler 1988) and it is post-Danian in age due to its unconformable boundary with the underlying Sobral Formation. The medium to coarse-grained, bluff-

Fig. 3.2 Landscape of the northern end of Seymour Island, Antarctic Peninsula showing the continuous surface exposure of Campamento, *Cucullaea* I, *Cucullaea* II and Submeseta allomembers of La Meseta Formation (looking west)

forming volcaniclastic sandstones of the Wiman member transitionally covers the former and comprises Sadler (1988) Tps5 unit, the "Wiman Formation" (Elliot and Hoffman 1989) beds and the Tpcv 3-4, being dated as early Upper Paleocene (Askin 1988). The upper Bahía Pingüino member rests on an erosional surface draped with slide-blocks and covered with dark, homogeneous mudstones with scattered and poorly preserved marine fossil remains like shark and fish teeth, gastropods, bivalves and crinoids. The fossil penguin bones described in this paper were unearthed in situ from a single level a few meters from the base of this unit. This lithofacies is in turn sharply overlain by an alternation of medium to fine-grained sandstones and mudstones containing a few pieces of silicified tree trunks and well preserved plant-leaf fossils (Dusén 1908) dated as Upper Paleocene. The lower part (Cerro Arañado and Wiman members), consisting of coarse-grained volcaniclastics, is due mainly to axial mass-flow deposits (Doktor et al. 1988; Elliot 1995). The upper third (Bahía Pingüino member) is made up of fine to very fine siliciclastic sandstones and mud-stones, and represents sedimentation in very shallow marginal marine to transitional (deltaic) environments developed after most of the relief was subdued.

3.2.2 La Meseta Formation

The early Eocene to ?earliest Oligocene La Meseta Formation (Elliot and Trautman 1982; Ivany et al. 2006) is an unconformity-bounded unit (La Meseta Alloformation of Marenssi et al. 1996; 1998a) (Fig. 3.1). This unit has a maxi-mum composite thickness of 720 meters filling up a 7 km wide valley cut down into the older sedimentary rocks of the island after the regional uplift and tilting of the Paleocene and Marambio Group beds. The La Meseta Formation comprises mostly poorly consolidated siliciclastic fine-grained sediments deposited in deltaic, estuarine and shallow marine environments as a part of a tectonically-controlled incised valley system (Marenssi 1995; Marenssi et al. 1998b), which spans nearly all of the Eocene, and for some authors includes the

Eocene–Oligocene boundary (Ivany et al. 2006). The age of the La Meseta Formation has received much attention from numerous authors. Montes et al. (2010) presented an up-to-date chronostratigraphic synthesis and provided new and independent age constrains for the age of the La Meseta Formation and its internal units which shall be used in this paper.

The La Meseta Formation was subdivided by Sadler (1988) into seven lithofacies units (Telm1-7). These were later organized into six erosionally-based internal units, named from base to top Valle de Las Focas, Acantilados, Campamento, *Cucullaea* I, *Cucullaea* II and Submeseta Allomembers (Marenssi et al. 1998a) (Fig. 3.2). These lens-shaped unconformity-bounded units are thought to represent different sedimentation stages related to sea-level fluctuations (Marenssi et al. 2002; Marenssi 2006). Marenssi et al. (1998a) recognized three main facies associations in the La Meseta Formation. They were deposited in deltaic, estuarine and shallow marine settings, mostly within a northwest-southeast trending valley (Marenssi et al. 1998a, b). Sedimentary environments such as tidal channels and flats, an estuary mouth platform, and a mid-estuary (Marenssi 1995; Marenssi et al. 1998a) formed a coastal area of low relief. In contrast, far inland to the west, the Antarctic Peninsula was a highland or mountainous area characterized by volcanoes sporadically active since the Mesozoic. Although paleogeographical interpretations indicate that terrestrial facies had to be present nearby to the west of Seymour Island, they are not yet known; hence, all terrestrial fossils reported to date are considered to have been transported within marine settings. Land-derived fossils were concentrated in paralic and shallow marine environments after some transport. The presence of leaves, tree trunks, and a flower suggest a forested terrain nearby (Gandolfo et al. 1998a, b; Torres et al. 1994; Doktor et al. 1996).

Provenance studies on sandstones of the La Meseta Formation demonstrate that the sediments came from the west-northwest, the source rocks being those outcropping on the Antarctic Peninsula (Marenssi 1995; Marenssi et al. 2002; Net and Marenssi 1999). Additionally, paleocurrent measurements confirm the location of the source area (Marenssi 1995). Therefore, the source area of the sediments, leaves and trunks was the northern Antarctic Peninsula, a magmatic arc that underwent uplift during the Cretaceous and Cenozoic (Elliot 1988). This cordillera supported forests in a range of habitats from coastal to alpine. Seymour Island lies on the eastern (back-arc) margin of the Antarctic Peninsula.

References

Askin RA (1988) Campanian to Paleocene palynological succession of Seymour and adjacent islands, northeast Antarctic Peninsula. In: Feldmann RM, Woodburne MO (eds), Geology and paleontology of Seymour Island, Antarctic Peninsula: Boulder, Colorado, Geological Society of America Memoir 169:131–153

Birkenmajer K (1995) Mesozoic-Cenozoic magmatic arcs of Northern Antarctic Peninsula: Subduction, Rifting and Structural Evolution. In: Srivastava RK, Chandra R (eds) Magmatism in relation to diverse tectonic settings. Oxford & IBH Publishing Co. PVT. LTD, New Delhi, pp 329–344

Buatois LA, López-Angriman AO (1992) Trazas fósiles y sistemas deposicionales, Grupo Gustav, Cretácico de la isla James Ross. In: Rinaldi CA (ed) Geología de la Isla James Ross. Antártida, Instituto Antártico Argentino, Buenos Aires, pp 239–262

Crame JA, Pirrie D, Riding JB, Thomson MRA (1991) Campanian-Maastrichtian (Cretaceous) stratigraphy of the James Ross Island area, Antarctica. J Geol Soc Lond 148:1125–1140. doi:10.1144/gsjgs.148.6.1125

del Valle RA, Elliot DH, Macdonald DIM (1992) Sedimentary basins on the east flank of the Antarctic Peninsula: proposed nomenclature. Antarct Sci 4:477–478

Doktor M, Gazdzicki A, Marenssi SA, Porebski SJ, Santillana SN, Vrba AV (1988) Argentine–Polish geological investigations on Seymour (Marambio) Island, Antarctica, 1988. Polish Polar Research 9:521–541

Doktor M, Gazdzicki A, Jermanska A, Porebski SJ, Zastawniak E (1996) A plant-and-fish assemblage from the Eocene La Meseta formation of Seymour Island (Antarctic Peninsula) and its environmental implications. Palaeontologia Polonica 55:127–146

Dusén P (1908) Über die tertiäre flora der Seymour-Insel. Wissenschaftliche Ergebnisse der Schwedischen Südpolar-Expedition 1901–1903, Lithographisches Institut des Generalstabs. Stokholm, Bd. 3, l. 3, 4 tf, pp 127

Dutra TL, Batten D (2000) The Upper Cretaceous flora from King George Island, an update of information and the paleobiogeographic value. Cretac Res 21(2–3):181–209

Elliot DH, Trautman TA (1982) Lower tertiary strata on Seymour Island, Antarctic Peninsula. In: Craddock C (ed) Antarctic geoscience. University of Wisconsin Press, Madison, pp 287–297

Elliot DH (1988) Tectonic setting and evolution of the James Ross Basin, northern Antarctic Peninsula. In: Feldmann RM, Woodburne MO (eds) Geology and Paleontology of Seymour Island, Antarctic Peninsula. Geological Society of America, Memoir 169:541–555

Elliot DH (1995) Paleogene volcanic rocks on Seymour Island: evidence for northwestward subduction of Weddell Sea oceanic crust. Abstract, VII International Symposium on Antarctic Earth Sciences, Siena, p.119

Elliot DH, Hoffman SM (1989) Geologic studies on Seymour Island. Antarctic Journal of the United States, XXIV(5):3-5

Gandolfo MA, Marenssi SA, Santillana SN (1998a) Flora y paleoclima de la Formación La Meseta (Eoceno medio), isla Marambio (Seymour), Antártida. In: Casadio S (ed) Paleógeno de América del Sur y de la Península Antártica. Asociación Paleontológica Argentina, vol 5. Publicación Especial, pp 155–162

Gandolfo MA, Hoc P, Santillana S, Marenssi S (1998b) Una flor fósil morfológicamente afín a las Grossulariaceae (Orden Rosales) de la Formación La Meseta (Eoceno medio), Isla Marambio, Antártida. In: Casadio S (ed) Paleógeno de América del Sur y de la Península Antártica. Asociación Paleontológica Argentina, vol 5. Publicación Especial, pp 147–153

Hathway B (2000) Continental rift to back-arc basin: Jurassic–Cretaceous stratigraphical and structural evolution of the Larsen Basin, Antarctic Peninsula. J Geol Soc Lond 157:417–432

Ineson JR (1989) Coarse-grained submarine fan and slope apron deposits in a Cretaceous back-arc basin, Antarctica. Sedimentology 36:739–819

Ivany LC, Van Simaeys S, Domack EW, Samson SD (2006) Evidence for an earliest Oligocene ice sheet on the Antarctic Peninsula. Geology 34(5):377–380

Marenssi SA (1995) Sedimentología y paleoambientes de sedimentación de la Formación La Meseta, isla Marambio, Antártida. Tomo I, pp 330, Tomo II, pp 172. Ph.D. dissertation, Universidad de Buenos Aires (unpublished)

Marenssi SA (2006) Eustatically-controlled sedimentation recorded by Eocene strata of the James Ross Basin, Antarctica. In: Francis JE, Pirrie D, Crame JA (eds) Cretaceous-Tertiary high-latitude palaeoenvironments, James Ross Basin, Antarctica, vol 258. Geological Society of London, Special Publications, pp 125–133

Marenssi SA, Santillana SN, Rinaldi CA (1996) Stratigraphy of La Meseta Formation (Eocene), Marambio Island, Antarctica. I Congreso Paleógeno de América del Sur, Santa Rosa, La Pampa. Abstracts Volume, pp. 33-34

Marenssi SA, Santillana SN, Rinaldi CA (1998a) Paleoambientes sedimentarios de la Aloformación La Meseta (Eoceno), Isla Marambio (Seymour), Antártida, vol 464. Instituto Antártico Argentino, Contribución, pp 51

Marenssi SA, Santillana SN, Rinaldi CA (1998b) Stratigraphy of the La Meseta Formation (Eocene), Marambio (Seymour) Island, Antarctica. In: Casadio S (ed) Paleógeno de América del Sur y de la Península Antártica. Asociación Paleontológica Argentina, Publicación Especial, vol 5. pp 137–146. Buenos Aires

Marenssi SA, Net LI, Santillana SN (2002) Provenance, depositional and paleogeographic controls on sandstone composition in an incised valley system: the Eocene La Meseta Formation, Seymour Island Antarctica. Sediment Geol 150(3–4):301–321

Marenssi SA, Santillana SN, Net LI, Rinaldi CA (2012) Heavy mineral suites as provenance indicator: La Meseta Formation (Eocene), Antarctic Peninsula. Asoc. Sedimentol. Argent 5:9–19

Montes M, Nozal F, Santillana S, Tortosa F, Beamud E, Marenssi S (2010) Integrate stratigraphy of the Upper Paleocene-Eocene strata of Marambio (Seymour) Island, Antarctic Peninsula. XXXI SCAR, Open Science Conference, Buenos Aires, Argentina

Net LI, Marenssi SA (1999) Petrografía de las areniscas de la Formación La Meseta (Eoceno), isla Marambio, Antártida. IV Jornadas sobre Investigaciones Antárticas, Buenos Aires, vol 2. pp 343–347

Olivero EB, Scasso RA, Rinaldi CA (1986) Revision of the Marambio Group, James Ross Island, Antarctica. Contribuciones Cientificas del Instituto Antártico Argentino 331:1–28

Olivero EB, Martinioni DR, Mussel FJ (1992) Upper Cretaceous sedimentology and biostratigraphy of western Cape Lamb (Vega Island, Antarctica). Implications on sedimentary cycles and evolution of the basin. In: Rinaldi CA (ed), Geología de la Isla James Ross. Instituto Antártico Argentino, Buenos Aires pp. 147–166

Olivero EB (2012) Sedimentary cycles, ammonite diversity and palaeoenvironmental changes in the Upper Cretaceous Marambio Group, Antarctica. Cretaceous Research 34:348–366. doi:10.1016/j.cretres.2011.11.015

Pirrie D, Crame JA, Riding JB (1991) Late Cretaceous stratigraphy and sedimentology of Cape Lamb, Vega Island, Antarctica. Cretaceous Research 12:227–258

Pirrie D, Crame JA, Lomas SA (1997) Late Cretaceous stratigraphy of the Admiralty Sound region, James Ross Basin, Antarctica. Cretaceous Research 18:109–137

Riding JB, Crame JA (2002) Aptian to Coniacian (early-late cretaceous) palynostratigraphy of the Gustav Group, James Ross Basin, Antarctica. Cretac Res 23:739–760

Rinaldi CA (1982) The upper cretaceous in the James Ross Island group. In: Craddock C (ed) Antarctic geoscience. The University of Wisconsin Press, Madison, pp 331–337

Sadler P (1988) Geometry and stratification of uppermost Cretaceous and Paleogene units of Seymour Island, northern Antarctic Peninsula. In: Feldmann RM, Woodburne MO (eds) Geology and Paleontology of Seymour Island, Antarctic Peninsula, vol 169. Geological Society of America, Memoir, pp 303–320

Torres T, Marenssi SA, Santillana SN (1994) Maderas fósiles de la isla Seymour, Formación La Meseta, Antártica. Serie Científica del INACH 44:17–38

Whitham AG, Ineson JR, Pirrie D (2006) Marine volcaniclastics of the Hidden Lake Formation (Coniacian) of James Ross Island, Antarctica: an enigmatic element in the history of a backarc basin. In: Francis JE, Pirrie D, Crame JA (eds), Cretaceous-Tertiary High-Latitude Palaeoenvironments, James Ross Basin, Antarctica. Geological Society, London, Special Publication 258:2–47

Wrenn JH, Hart GF (1988) Paleogene dinoflagellates cyst biostratigraphy of Seymour Island, Antarctica. In: Feldmann RM, Woodburne MO (eds) Geology and Paleontology of Seymour Island, Antarctic Peninsula, vol 169. Geological Society of America, Memoir, pp 321–447

Chapter 4
South America/West Antarctica: Pacific Affinities of the Weddellian Marine/Coastal Vertebrates

The high latitude Weddellian Biogeographical Province, conceived by Zinsmeister (1979, 1982) on the basis of marine molluscan, echinoderm and arthropod faunas, was defined as a cool temperate, shallow water region which extended from southern South America (Magallanic Region in Chile, and Tierra del Fuego and Santa Cruz provinces in Argentina), along the Antarctic Peninsula and West Antarctica, to New Zealand, Tasmania and southeastern Australia (Fig. 4.1). Case (1988) expanded this concept to a "biogeographic province" with the inclusion of terrestrial plants and mammals. This province existed from the Late Cretaceous through the Eocene when Australia, Antarctica, and southernmost South America were in proximity (Zinsmeister 1979, 1982; Woodburne and Zinsmeister 1984).

The final early Paleogene isolation of South America probably did not occur until a deep-water seaway developed between southern South America and the Antarctic Peninsula. This seaway was the main factor that helped to preserve the ecological affinities of marine vertebrate assemblages distributed along different parts of the Weddellian Province (Zinsmeister 1979), while endemic assemblages seem to have been restricted to shallower waters, suggesting that these evolved as a result of local biogeographic barriers. Opening of the Drake Passage gateway between the Pacific and Atlantic oceans has been linked in various ways to Cenozoic climate changes. All but one of the available constraints on the age of the central Scotia Sea is diagnostic of, or consistent with, a Mesozoic age. Comparison of tectonic and magnetic features on the seafloor with plate kinematic models shows that it is likely to have accreted to a mid-ocean ridge between the South American and Antarctic plates following their separation in Jurassic times.

Before final fragmentation of western Gondwana during the early Paleogene, the southern circum-Pacific closely resembled the present-day northern Pacific setting, with nearly continuous land extending across the pole from South America to Australia. The southern margin of the Pacific was isolated by both geography and oceanic circulation from the other major oceans of the Southern Hemisphere (Zinsmeister 1982). The marine Middle Eocene deposits extend into the Austral Basin, cropping out both in Tierra del Fuego and Santa Cruz. Dinocyst assemblages from northeastern

M. Reguero et al., *Late Cretaceous/Paleogene West Antarctica Terrestrial Biota and Its Intercontinental Affinities*, SpringerBriefs in Earth System Sciences, DOI: 10.1007/978-94-007-5491-1_4, © The Author(s) 2013

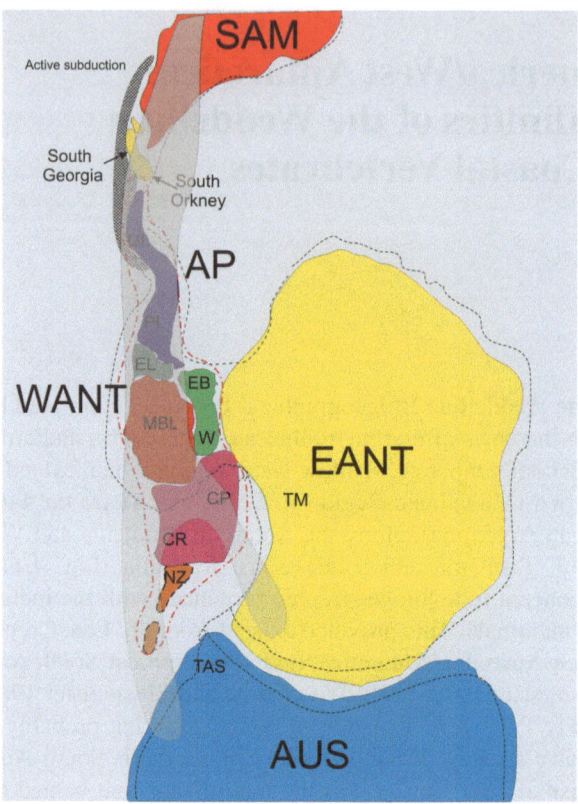

Fig. 4.1 Sketch map of the Late Cretaceous paleogeography of the Weddellian Province (after Woodburne and Zinsmeister 1984). *AP*, Antarctic Peninsula (West Antarctica); *CP*, Campbell Plateau; *CR*, Chatham Rise; *EANT*, East Antarctica; *EB*, Ellsworth Block; *EL*, Ellsworth Land; *GL*, Graham Land; *MBL*, Marie Byrd Land; *NZ*, New Zealand; *PL*, Palmer Land; *TA*, Tasman Rise; *TM*, Transantarctic Mountains; *W*, Whitmore Mountains; *WANT*, West Antarctica

Tierra del Fuego (Leticia Formation) are dominated by the so-called circum-Antarctic, endemic, Transantarctic Flora. The Transantarctic Flora is also recorded from southwestern Santa Cruz within the Río Turbio Formation. Major faunistic marine/coastal vertebrate affinities between South America and West Antarctica come from the Magallanes Region in the southern part of Chile. Sallaberry et al. (2010) described several localities that have yielded marine fossil vertebrates with clear Weddellian affinities. One of these localities is Río de las Minas, located 10 km west from Punta Arenas. The host beds outcrop on the upper slopes of the Las Minas River valley and forms part of the Loreto Formation to the Eocene-Miocene. The Loreto Formation is about 800 m thick, and it is composed of well-sorted, marine sandstones containing glauconitic and concretionary horizons, intercalated with beds containing fossil flora and coal seams with variable thickness. The fossil content of bivalves, gastropods and vertebrates indicates that this unit was deposited in a marine environment, although wood fragments and leaf imprints are also present. From the same stratigraphic level of this unit were recovered abundant elasmobranch teeth predominantly belonging to the species *Striatolamia macrota* and dental plates of the chimaeroid *Ischyodus dolloi*. *S. macrota* was mentioned by Kriwet (2005) as the predominant taxon of elasmobranchs in all the strata and associations of the La Meseta Formation. These taxa suggest an Early Eocene age for the host beds. This temporal assignment is consistent with the Early Eocene age assigned to the ichthyofauna found in the middle levels (*Cucullaea* I Allomember) of the La Meseta Formation in Seymour Island (Cione and Reguero 1994, 1998; Long 1992a; Reguero et al. submitted).

The fossiliferous levels of one of these localities, Sierra Dorotea, are located near the border between Chile and Argentina and belong to a Chilean unit equivalent to the Río Turbio Formation. The Río Turbio Formation is formed by sandstones, conglomerates and coquina horizons with intercalated continental clays. Its fossil content includes the bivalves *Glycimeris* cf. *G. ibari*, *Panopea* cf. *P. clausa* and *Venericardia* (*Venericor*) *carrerensis* all previously recognized by Griffin (1991) in outcrops of the Río Turbio Formation exposed in Argentina and assigned to the Eocene. The presence of fossil penguins of the genus *Palaeeudyptes* associated with elasmobranch teeth of the species *Striatolamia macrota* and abundant teeth of *Carcharias* sp. allow these beds to be correlated with outcrops of the La Meseta Formation of Seymour Island.

4.1 Late Cretaceous/Paleogene Marine Fossil Vertebrates of the James Ross Basin

The Antarctic Late Cretaceous/Paleogene marine vertebrates from the James Ross Basin are numerous and diverse (Reguero and Gasparini 2007), and include plesiosaurs, mosasaurs, sharks, bony fishes, turtles, penguins and whales. We comment here on the known record of sharks, bony fishes, turtles, and whales. The penguin record (Sphenisciformes) is considered on the next section of this work.

4.1.1 *Neoselachii and Teleostei Fossil Fishes*

Fossil fish remains have long been known from the extensive Cretaceous marine deposits of the James Ross Basin on the NE flank of the Antarctic Peninsula. The first fossils were collected during the 1901–1903 Swedish South Polar Expedition and were subsequently described by Woodward (1908). Selachian remains are the main component of the Late Cretaceous marine vertebrate fauna. Sharks include at least 11 species of Hexanchiformes, Lamniformes, Squatiniformes and Synechodontiformes (Kriwet et al. 2006, see Table 4.1).

The telesotean fauna of the Late Cretaceous of the James Ross Basin is rather low in diversity compared to contemporaneous fish faunas (Table 4.1). Rays, an important component of marine fish associations, as well as fish from lower trophic levels, remain unknown from the Late Cretaceous of Antarctica.

The Antarctic Eocene marine vertebrates from the James Ross Basin are numerous and diverse (Reguero and Gasparini 2007), and they come exclusively from La Meseta Formation, Seymour Island.

Table 4.1 Taxonomic list, stratigraphy, geographic locations and references for Chondrichthyes and Teleostei fishes from the Late Cretaceous of the James Ross Basin, Antarctic Peninsula

Taxon	Stratigraphy (Formation)	Geographic location	Source
Chondrichthyes			
Fam. Clamydoselachidae			
Clamydoselachus thomsoni	SM	James Ross	Richter and Ward (1990)
Fam. Odontaspidae			
Odontaspis	LDB	Seymour	Martin and Crame (2006)
Fam. Hexanchidae			
Notidanodon dentatus	LDB, SM	Seymour, James Ross, Vega	Cione and Medina (1987)
Fam. Squatinidae			
Squatina sp.	SM	James Ross	Richter and Ward (1990)
Fam. Orthacodontidae			
Sphenodus sp.	SM, LDB	Seymour and James Ross	Grande and Eastman (1986)
Fam. Palaeospinacidae			
Paraorthacodus sp.	SM	James Ross	Kriwet et al. (2006)
Fam. Mitsukurinidae			
Scapanorhynchus sp.	SM	James Ross	Kriwet et al. (2006)
Selachii indet.			
"*Ptychodus*"	LDB	Seymour	Woodward (1908)
Holocephali			
Fam. Callorhynchidae			
Ischyodus dolloi	LDB	Seymour	Stahl and Chatterjee (2002)
Fam. Chimaeridae			
gen. et sp. indet.	LDB	Seymour	Stahl and Chatterjee (1999)
Chimaera zangerli	SM, LDB	James Ross, Seymour	Kriwet et al. (2006)
Teleostei			
Beryciformes			
Antarctiberyx seymouri	LDB	Seymour	Grande and Chatterjee (1987)
Ichthyodectiformes indet.	SM	James Ross	Kriwet et al. (2006)
Albuliformes indet.	SM	James Ross	Kriwet et al. (2006)
Fam. Enchodontidae			
Enchodus sp.	SM	James Ross	Richter and Ward (1990)
Apateodus? sp.	SM	James Ross	Kriwet et al. (2006)
Actinopterygii indet.			
Sphaeoronodus? sp.	SM	James Ross	Richter and Ward (1990)
cf. Sphenocephalidae	LDB	Vega	Martin and Crame (2006)

Abbreviations LDB, López de Bertodano, *HL, SM*, Santa Marta

The Eocene selachian fauna from Antarctica includes 25 species in 16 families (Table 4.2). 24 taxa and 15 families come from the *Cucullaea* I Allomember of the La Meseta Formation of Seymour Island. The majority of taxa belong to sharks

Table 4.2 Taxonomic list, stratigraphy, and references for the Chondrichthyes fauna from the La Meseta Formation of Seymour Island, Antarctic Peninsula

Taxon	Stratigraphy (Allomember)	Source
Chondrichthyes		
Fam. Hexanchidae		
Heptranchias howelli	*Cucullaea* I	Long (1992a)
Hexanchus sp.	*Cucullaea* I	Cione and Reguero (1994)
Fam. Squalidae		
Squalus woodburnei	*Cucullaea* I	Long (1992a)
Squalus weltoni	*Cucullaea* I	Long (1992a)
Centrophorus sp.	*Cucullaea* I	Long (1992a)
Dalatias licha	*Cucullaea* I	Long (1992a)
Deania sp.	*Cucullaea* I	Long (1992a)
Fam. Squatinidae		
Squatina sp.	*Cucullaea* I	Welton and Zinsmeister (1980)
Fam. Pristiophoridae		
Pristiophorus lanceolatus	*Cucullaea* I, Submeseta	Grande and Eastman (1986)
Fam. Ginglymostomatidae		
Pseudoginglymostoma cf. *P. brevicaudatum*	*Cucullaea* I	Long (1992a)
Fam. Orectolobidae		
Stegostoma cf. *S. fasciatum*	*Cucullaea* I	Long (1992a)
Fam. Odontaspididae		
Palaeohypototus rutoti	Acantilados, *Cucullaea* I	Long (1992a)
Odontaspis winkleri	Acantilados, *Cucullaea* I	Long (1992a)
Striatolamia macrota	Acantilados, *Cucullaea* I	Kriwet (2005)
Fam. Cetorhinidae		
Cetorhinus sp.	*Cucullaea* I	Cione and Reguero (1998)
Fam. Lamnidae		
Isurus praecursor	*Cucullaea* I	Cione and Reguero (1994)
Lamna cf. *L. nasus*	*Cucullaea* I	Long (1992a)
Fam. Otodontidae		
Carcharocles auriculatus	*Cucullaea* I, Submeseta	Welton and Zinsmeister (1986)
Fam. Mitsukurinidae		
Anomotodon multidenticulata	*Cucullaea* I	Long (1992a)
Fam. Carcharhinidae		
Scoliodon sp.	*Cucullaea* I	Long (1992a)
Carcharhinus sp.	*Cucullaea* I	Kriwet (2005)
Fam. Palaeospinacidae		
Paraorthacodus sp.	*Cucullaea* I	Cione (comm. pers.)
Fam. Pristidae		
Pristis sp.	*Cucullaea* I	Kriwet (2005)
Fam. Myliobatidae		
Myliobatis sp.	Acantilados, *Cucullaea* I	Cione et al. (1977)
Fam. Rajidae		

(Continued)

Table 4.2 (Continued)

Taxon	Stratigraphy (Allomember)	Source
Bathyraja sp.	*Cucullaea* I	Long (1992c)
Holocephali		
Fam. Callorhynchidae		
Ischyodus dolloi	*Cucullaea* I	Grande and Eastman (1986)
Fam. Chimaeridae		
Chimaera seymouriensis	*Cucullaea* I, Submeseta	Ward and Grande (1991)

while batoids are represented by only three taxa with a very uneven distribution in the sequence. Long and Stilwell (2000) reported the presence of rare selachian teeth from Eocene deposits of Mount Discovery in East Antarctica. This material includes the first record of *Galeorhinus* for Antarctica.

Sharks remains, largely represented by isolated teeth and vertebrae and a few poorly preserved dorsal spines, are extremely abundant and diverse in the coarser sand facies of the *Cucullaea* I Allomember. The *Cucullaea* I Allomember contains the bulk of the fossil shark's localities. During the austral summers of 1992–2000 more than 10,000 teeth of fishes were recovered by Argentinean teams by dry-sieving and surface-prospecting from four localities (DPV 2/84, DPV 6/84, IAA 1/90 and 2/95, Fig. 1.3) of the *Cucullaea* I Allomember. Interestingly the level of diversity from a single locality (IAA 1/90) of the La Meseta Formation is much higher than the level of diversity for most extant cool temperate shark faunas and nearly equal to a present-day tropical shark fauna. At least 21 taxa of sharks representing 11 families (Table 4.2) occur in this horizon. Long (1992a) pointed out that this Antarctic fossil shark assemblage constitutes a very complex assortment of sharks from many different habitats, converging on a single thanatocenosis.

We only considered neoselachian proportions from the best sampled localities (IAA 1/90; and in part IAA 2/95) which correspond to a bank of naticids (mostly composed by the gastropod *Polynices*) in the *Cucullaea* I Allomember. The most abundant (according to tooth number) elamosmobranch taxa in this horizon are *Squatina* (37.89 %), *Pristiophorus* (22.45 %), Odontaspididae (17.24 %), *Myliobatis* (6.70 %), *Squalus* (6.81 %), Rajidae (4.72) and Holocephali (2.68 %)(Cione et al. 2007). The rare occurrence of *Carcharhinus* and *Pristis*, both taxa confined today to warm-temperate to tropical waters, in the temperate waters of Early Eocene in Antarctica indicates that both were not primary inhabitants but instead migrated along open trans-equatorial seaways into Southern Hemisphere waters (Kriwet 2005).

Most of the fish taxa mentioned here are found in many different overlying localities and horizons of the *Cucullaea* I Allomember except in the upper units, which document a sharp decrease in diversity near the boundary *Cucullaea* II and Submeseta allomembers. In the Submeseta Allomember, there is a dramatic diminution of diversity of those selachians that dominated below; there are no teleost taxa characteristic of warm water (e.g., Labridae, Oplegnathidae, Xiphiorhynchidae); sharks dramatically decrease in diversity and quantity (a few *Pristiophorus* and odontaspidid teeth) and begin to feature some sharks (*Lamna*)

Table 4.3 Taxonomic list, stratigraphy, and references for the teleostean fishes and reptiles from the La Meseta Formation of Seymour Island, Antarctic Peninsula

Taxon	Stratigraphy (Allomember)	Source
Clupeiformes		
Fam. Clupeidae		
Marambionella andreae	Acantilados, *Cucullaea* I?	Jerzmanska (1991)
Perciformes		
Fam. Oplegnathidae		
Oplegnathus sp.	*Cucullaea* I	Cione et al. (1994)
Fam. Xiphiorhynchidae		
cf. *Xiphiorhynchus*	Campamento, *Cucullaea* I	Cione et al. (2001)
Fam. Trichiuridae		
Trichiurus sp.	*Cucullaea* I	Long (1991)
Fam. Labridae		
gen. et sp. indet.	*Cucullaea* I	Long(1992b)
Siluriformes		
Incertae sedis		
Siluriformes undetermined	*Cucullaea* I?	Grande and Eastman (1986)
Gadiformes		
Fam. Merluccidae		
"*Mesetaichthys*"	*Cucullaea* I, Submeseta	Jerzmanska and Swidnicki (1992)
Beryciformes		
gen. et sp. indet.	Acantilados, *Cucullaea* I?	Doktor et al. (1996)
Cryptodira		
Fam. Dermochelyidae		
"*Psephophorus*"sp.	*Cucullaea* I	De la Fuente et al. (1995)
Testudines indet.	*Cucullaea* I	Bona et al. (2010)

and teleosts with species characteristic of colder waters (e.g., gadiforms of the informal genus "*Mesetaichthys*") are recorded.

Long (1992c) and Case (1992) analyzed the ecology and diversity of the Eocene Seymour selachian fauna and concluded that the selachian fauna represents a cool-temperate fauna with different ecological components including tropical water immigrants (e.g., *Pseudoginglymostoma*, *Stegostoma*, *Scoliodon*).

While the diversity of the elasmobranch fauna in the *Cucullaea* I Allomember of the La Meseta Formation seems to be abundant and quite diverse, the teleost fishes appear to be scant and poor in diversity. Notwithstanding, teleost fishes are represented by gadiforms, clupeiforms, oplegnathids, siluriforms, perciforms, beryciforms, and chimaeriforms (see Table 4.3).

A large osteichthyan vertebra tentatively assigned to the rare genus *Xiphiorhynchus* (billfish) was recovered from the Acantilados Allomember (Cione et al. 2001). According to strontium (Sr) isotope dating (Reguero et al. 2002), the age of the fossil-bearing horizon is between 52.4 and 54.3 Ma. The presence of billfish in Antarctica agrees with the high temperature suggested for the time of the deposition of the fossiliferous rocks.

4.1.2 Marine Reptiles

Evidences of Late Cretaceous (Coniacian to Maastrichtian) Antarctic marine
reptiles are documented exclusively in the NE flank of the Antarctic Peninsula,
West Antarctica (Reguero and Gasparini 2007). Coniacian, Campanian and
Maastrichtian deposits of the James Ross Basin have been extensively surveyed
by vertebrate paleontologists; most collections consist of marine reptiles (del
Valle et al. 1977; Gasparini et al. 1984; Chatterjee and Small 1989; Martin et al.
2002; Novas et al. 2002; Martin 2006; Martin and Crame 2006; de la Fuente et
al. 2010) and other marine vertebrates (Woodward 1908; Cione and Medina
1987; Grande and Chatterjee 1987; Richter and Ward 1990; Stahl and Chatterjee
1999, 2002; Kriwet et al. 2003, 2006). Marine reptiles were some of the first
vertebrate fossils to be found in the Late Cretaceous of Antarctica, plesiosaurs and
the marine squamates, mosasaurs, were initially reported by del Valle et al. (1977)
and Gasparini and del Valle (1981). Since then, other expeditions have found limited and fragmentary plesiosaur specimens on the islands of that same area. All
those occurrences are referred to the Elasmosauridae, to *Aristonectes* (regarded as
an Elasmosauridae or an Aristonectidae) or as Plesiosauria indet. (Gasparini et al.
1984; Chatterjee and Small 1989; Fostowicz-Frelic and Gazdzicki 2001; Martin
and Crame 2006; Martin et al. 2007). Most of these came from Seymour and Vega
islands, from the López de Bertodano and Snow Hill Island formations, of the
Marambio Group (Table 4.4).

Kellner et al. (2011) described remains of the oldest plesiosaur (Santonian,
Lachman Crags Member, Santa Marta Formation, James Ross Island) known so
far from Antarctica and the first one not referable to the Elasmosauridae nor to
Aristonectes from this region, indicating a higher diversity of this group of aquatic
reptiles in this continent than previously suspected.

De la Fuente et al. (2010) described a partially preserved chelonioid carapace
found in the Santa Marta Fm., James Ross Island, thus constituting the oldest
known turtle from Antarctica. The new discovery enlarges the meager fossil record
of turtles from this continent.

The record of marine reptiles in the Eocene of Antarctica is restricted to
turtles, a Cryptodira and a Testudines indet. (Table 4.3). De la Fuente et al. (1995)
assigned several plates from the *Cucullaea* I Allomember (Telm 4 of Sadler
1988) to a leatherback turtle (Dermochelyidae) very close to the extant species
Dermochelys coreacea. Albright et al. (2003) provisionally assigned these plates
to "*Psephophorus*" *terrypratchetti* Köhler, a species from upper Lutetian of South
Island, New Zealand.

Bona et al. (2010) describe two turtle carapace plates of a testudine recovered
from the middle levels (*Cucullaea* I Allomember) of the La Meseta Formation.
This material represents the first record of a turtle with a bony carapace from
the Eocene of Antarctica, and it increases the diversity of the group on this
continent.

Table 4.4 Taxonomic list, stratigraphy, geographic locations and references for marine and terrestrial reptiles from the Cretaceous of the James Ross Basin, Antarctic Peninsula

Taxon	Stratigraphy (Formation)	Geographic location	Source
Plesiosauria			
Elasmosauridae indet.	LDB	James Ross and Vega	del Valle et al. (1977)
cf. *Mauisaurus* sp.	LDB	Seymour	Chatterjee and Small (1989)
Aristonectes parvidens	LDB	Seymour	Gasparini et al. (2003)
gen. et sp. indet	SM	James Ross	Kellner et al. (2011)
Mosasauria			
Mosasauria indet.	LDB	Seymour.	Gasparini and del Valle (1981)
Tylosaurinae indet.	LDB	Seymour	Martin and Crame (2006)
Leiodon sp.	LDB	Vega, Seymour	Martin et al. (2002)
Pliopatecarpus sp.	LDB	Vega, Seymour	Martin et al. (2002)
cf. *Hainosaurus*	LDB	Vega	Martin et al. (2002)
Mosasaurus cf. *lemmonieri*	LDB	Vega, Seymour	Martin et al. (2002)
Taniwhasaurus antarcticus	SM	James Ross	Martin and Fernández (2005)
Mosasaurus sp.	LDB	Seymour	Martin and Crame (2006)
Dinosauria			
Ankylosauria			
Antarctopelta oliveroi	SM	James Ross	Salgado and Gasparini (2006)
Hypsilophodontidae	LDB	Vega	Hooker et al. (1991)
Fam. Hadrosauridae	LDB	Vega and James Ross	Case et al. (2000); Novas et al. (2002)
?Hadrosauridae indet.	LDB	Seymour	Rich et al. (1999)
Theropoda	HL	James Ross	Molnar et al. (1996)
Fam. Dromeosauridae			
gen. et sp. indet.	LDB	James Ross	Case et al. (2007)
Ornithopoda			
gen. et sp. nov.	SM	James Ross	Coria et al. (2008)
Dinosauria footprints	LDB	Snow Hill	Olivero (2007)
Testudines			
Chelionoidea *incertae sedis*	SM	James Ross	De la Fuente et al. (2010)

Abbreviations LDB López de Bertodano, *HL* Hidden Lake, *SM* Santa Marta

4.1.3 Whales

Whale remains occur sporadically throughout the La Meseta Formation (*Cucullaea* I and Submeseta allomembers). Archaeocetes from the La Meseta Formation of Seymour Island, Antarctica (Borsuk-Bialynicka 1988; Fostowicz-Frelik 2003), are in the main based on nondiagnostic postcranial material. Also Cozzuol (1988) reported the presence of the archaeocete *Zygorhiza* in the uppermost part of the La Meseta Formation (Submeseta Allomember). A primitive mysticete, *Llanocetus denticrenatus*, from the Submeseta

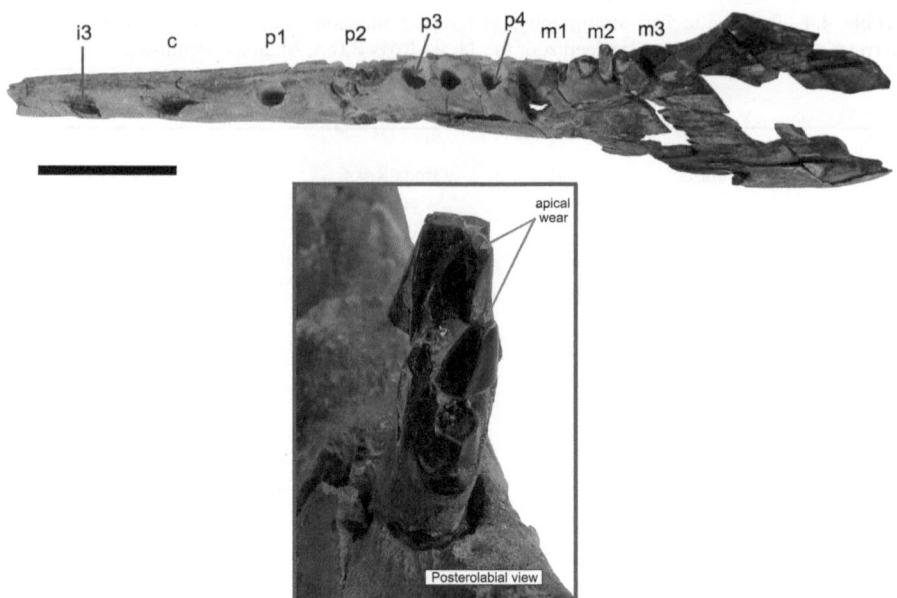

Fig. 4.2 Seymour Island Pelagiceti gen. et sp. nov., MLP 11-II-21-3, incomplete left dentary with p2 preserved in situ. Base of the *Cucullaea* I Allomember, the basal shell bed (Telm 4 of Sadler 1988) at DPV 2/84 locality. Scale bar = 5 cm

Allomember was described by Mitchell (1989). All supposed austral Archaeoceti, which include described specimens from Seymour Island, New Zealand, and Australia, are either indeterminate or are heterodont archaic Mysticeti (Fordyce and Barnes 1994).

Recently Reguero et al. (2011) reported an incomplete jaw with teeth (Fig. 4.2) of a new basilosaurid archaeocete recovered from the *Cucullaea* shell bank (Telm 4 of Sadler 1988) at DPV 2/84 locality (Fig. 4.3). This discovery represents the oldest basilosaurid known and increases the knowledge of the origin of the Pelagiceti clade providing new insights into the temporal evolution of fully aquatic whales.

The mandible bears five alveoli and two cheek teeth. The crown of one tooth, probably P2, is triangular and laterally compressed with multiple accessory denticles (3) and wear facet with a distinctive basal cingulid (Fig. 4.2).

4.2 Weddellian Sphenisciformes: Systematics, Stratigraphy, Biogeography and Phylogeny

The first collection of fossil penguin bones from Seymour Island was gathered by members of the Swedish South Polar Expedition in 1901–1903. The earliest published systematic of the extinct Sphenisciformes from that region (Wiman 1905a, b)

Fig. 4.3 Base of the
Cucullaea I Allomember,
the basal shell bed (Telm 4
of Sadler 1988) with very
abundant specimens of the
bivalve *Cucullaea raea* and
gastropod darwinellids at
DPV 2/84 locality, west side
of plateau at north end of
Seymour Island, Antarctic
Peninsula

distinguished six species of penguins (Fig. 4.4). Each of them was the type species
of a new genus.

Seymour Island spheniciformes are by far the dominant group of Eocene
marine/coastal vertebrates in Antarctica. After 30 years of paleontological investigations on Seymour Island, the penguin-bearing localities have increased
significantly with 20 localities containing penguin remains and constituting an
almost continuous record from the late Paleocene (Bahia Pingüino Allomember,
Cross Valley Formation, Thanetian, 55–56 Ma, Marenssi et al. 2012) to the latest Eocene (Submeseta Allomember, La Meseta Formation, Priabonian, 34 Ma,
Marenssi et al. 1998a). The bulk of the penguin-bearing sites is located within
the upper part of the Submeseta Allomember, in rocks belonging to the Facies
Association III of Marenssi et al. (1998a) (Fig. 4.5).

Clarke et al. (2003) proposed several phylogenetic group-definitions for higher
taxa within the Spheniciformes. Panspheniciformes is applied to the clade
including all taxa more closely related to Spheniscidae than any other extant
avian lineage. Spheniciformes is applied in a more exclusive sense to the clade
including all Panspheniciformes that share the apomorphic loss of aerial flight.
Spheniscidae is applied to the crown clade of penguins, comprising the most
recent common ancestor of all living penguin species and its descendants.

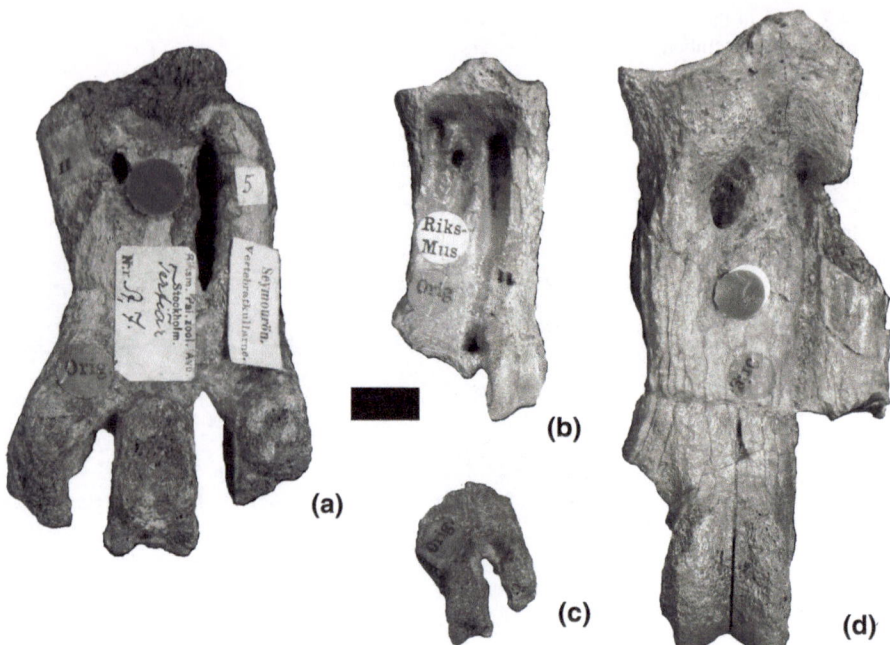

Fig. 4.4 Fossil penguins from Seymour Island, Antarctic Peninsula collected by the South Polar Swedish Expedition (1901–1903) and described by Wiman (1905a). **a** *Palaeeudyptes gunnari*, NRM A.7, holotype, incomplete tarsometarsus; **b** *Delphinornis larseni*, NRM A.21, holotype, incomplete tarsometatarsus; **c** *Ichtyopteryx gracilis*, NRM A.20, incomplete tarsometatarsus; **d** *Anthropornis nordenskjoeldi*, NRM A.45, holotype, incomplete tarsometatarsus. NRM (Naturhistoriska Riksmuseet of Stockholm). Scale bar = 5 cm

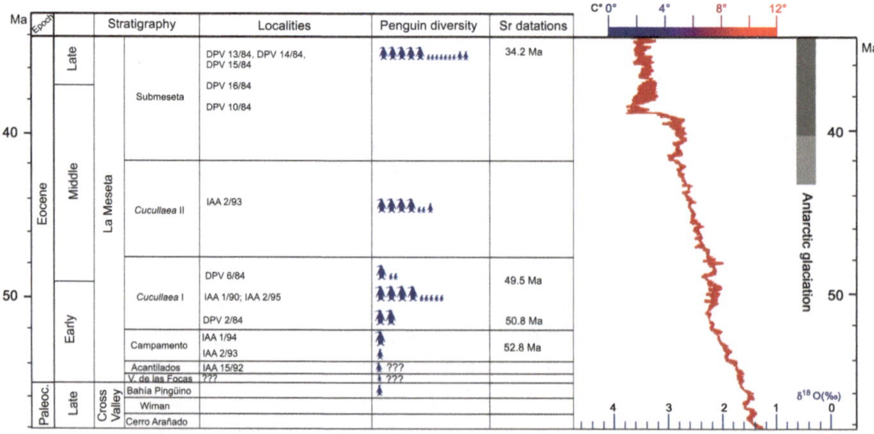

Fig. 4.5 Stratigraphical record of Seymour Island Sphenisciformes (Cross Valley and La Meseta formations)

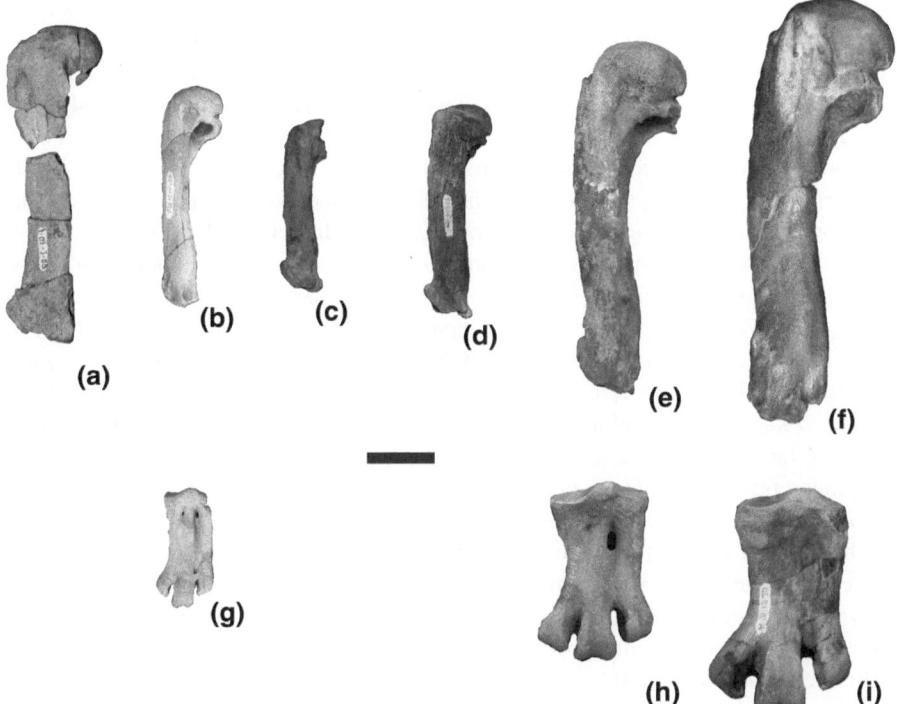

Fig. 4.6 Humeri of Weddellian penguins from the Paleogene of Seymour Island, Antarctic Peninsula. Comparative series in caudal view. Photographs have been reversed where necessary. **a** *Crossvallia unienwillia*, MLP 00-I-10-1, holotype, Cross Valley Formation (Paleocene), **b** *Delphinornis larseni*, MLP 93-X-1-146, La Meseta Formation (Eocene); **c** *Tonniornis mesetaensis*, MLP 93-X-1-145, holotype, La Meseta Formation (Eocene); **d** *Tonniornis minimun*, MLP 93-I-6-3, holotype, La Meseta Formation (Eocene); **e** *Palaeeudyptes gunnari*, MLP 93-X-1-3, La Meseta Formation (Eocene) and **f** *Anthropornis nordenskjoeldi*, MLP 93-X-1-4, La Meseta Formation (Eocene). Tarsometatarsi of Weddellian penguins from the Ypresian *Cucullaea* I (La Meseta Formation) Seymour Island, Antarctic Peninsula. Antarctica. Comparative series in cranial view; **g** *Delphinornis larseni*, MLP 93-X-1-161; **h** *Palaeeudyptes gunnari*, MLP 84-II-1-78 and **i** *Anthropornis nordenskjoeldi*, MLP 94-III-15-20. *MLP* Museo de La Plata, Argentina, Scale bar = 2 cm

The first collections of Antarctic fossil penguins, collected on Seymour Island by the Swedish expedition in 1901–1903, were studied by Wiman 1905a, b, who proposed a new approach to deal with the nature of fossil penguin assemblages. Due to the fragmentary state of the bones, he did not classify them all in a systematic scheme. Instead, he proposed grouping them in eight categories according to their size and robustness; the third, fifth and seventh of these show some degree of intra-group variability. Each category includes several remains that belong to different individuals but are morphologically similar and could be conspecific (Fig. 4.6).

Group number 1 was established for the largest synsacrum (holotype of *Orthopteryx gigas* Wiman 1905a, b), an element known only from a few fossil penguin species, and no other elements were subsequently assigned to it. It is large but not very robust, and slightly curved. Originally, Wiman (1905a, b) had certain doubts about the assignment of this synsacrum to a penguin, due to its size and particular morphological features. However, later revisions (Simpson 1946; see also Jadwiszczak 2009, Jadwiszczak and Mörs 2011) discarded that idea and considered it to belong to a sphenisciform. This taxon was considered as "essentially indeterminate" by Simpson (1971a) and recently, Jadwiszczak (2009) stated that this bone probably belonged to *A. nordenskjoeldi*. Jadwiszczak and Mörs (2011) carefully analyzed the morphology and size of this synsacrum, concluding that it should be in fact a synonyms of *Anthropornis nordenskjoeldi* (the latter having priority). However, a synsacrum is not a taxonomically diagnostic element. On this basis, and according to Art. 13 of the ICZN (1999), we regard *Orthopteryx gigas* as valid name but also a *nomen dubium* (Acosta Hospitaleche and Reguero 2011a).

Group number 2 remained unnamed and also consists of a synsacrum, but in this case incomplete. In a later revision, the sysnsacrum was re-examined, but its systematic position remains open to question (Jadwiszczak and Mörs 2011).

Group number 3 was identified as *Anthropornis nordenskjoeldii* Wiman 1905a, b on the basis of a tarsometatarsus and comprises several elements (tarsometatarsus, coracoid, humerus, ulna, carpometacarpus, femur, tibiotarsus, synsacrum) as well. However, subsequent revisions, particularly of the synsacrum, agree in their assignment to a smaller penguin, probably a species of *Palaeeudyptes* (Simpson, 1971; Jadwiszczak and Mörs 2011).

Group number 4, originally recognized as *Pachypteryx grandis* Wiman 1905a, b, was based on a tarsometatarsus and also contains others appendicular elements (coracoid, radius, carpometacarpus, tibiotarsus). In the current systematic scheme (Acosta Hospitaleche et al. 2007; Ksepka et al. 2006), this species is considered a synonym of *Anthropornis grandis* (Wiman 1905a, b). In a later revision, Simpson (1946) proposed that the four abovementioned groups be placed in the same category, probably assigned to *Anthropornis nordenskjoeldii*, an idea later supported by Marples (1953).

Group number 5 was associated with *Eospheniscus gunnari* Wiman 1905a, b— currently *Palaeeudyptes gunnari* (Wiman 1905a, b) Simpson 1971–on the basis of a tasometataarus, the type of the species. This group also comprises many other elements (coracoid, humerus, ulna, femur, tibiotarsus, synsacrum). It is worth repeating that none of these bones are associated and that they belong to different specimens. However, this was taken into account in subsequent works, and the allocations made by Wiman 1905a, b were considered in a systematic context.

Group number 6 was also innominate and was based on fragments of a humerus, coracoid, scapula, and femur. In a later review, *Notodyptes wimani* Marples 1953 was also included in this category. This species was erected on the basis of an incomplete tarsometatarsus from the same penguin assemblage in Seymour Island (Marples 1953) and placed later within *Archaeospheniscus wimani* (Marples 1953) Simpson (1971a).

Group number 7 includes a single tarsometatarsus that was the basis for the creation of *Delphinornis larsenii* Wiman (1905a, b).

Group number 8 represents the smallest species, identified as *Ichtyopteryx gracilis* Wiman (1905a, b). This species is based on a fragmentary tarsometatarsus described as *Ichthyopteryx gracilis* Wiman (1905a) by Simpson (1946) who subsequently considered it as a *nomen dubium* (Simpson (1971). Thereafter it was listed as a distinct species by Brodkorb (1963), although not considered as such by Marples (1953) due to its poor preservational state. Despite Simpson's assessment, this species was subsequently considered valid by several authors. For example, Bargo and Reguero (1998) listed *Ichthyopteryx gracilis* within the Argentinean Antarctic collections. More recently Myrcha et al. (2002) did not include these species in their Antarctic lists because the holotype described by Wiman (1905a, b) was too fragmentary to compare it with other penguins.

More recently, Jadwiszczak and Mörs (2011) analyzed the diversity in early Antarctic penguins, concluding that *Ichtyopteryx gracilis* should be considers a synonym of *Delphinornis gracilis* Myrcha et al. (2002). Their assumptions are based on *D. gracilis* and *I. gracilis* are closest to each other in terms of dimensions and also share a unique shape of the articular surface of the trochlea III in plantar view. They added the latter character to the revised diagnosis and established synonymy taking into account that *I. gracilis* has priority at specific level, and *Delphinornis* has it at generic level (besides, they would be secondarily homonyms). Here, we agree with Simpson (1971) that the lack of morphological details provides insufficient data to confidently discriminate the specimen as a distinct species. Its preservational state is inadequate to erect a new species. *Ichtyopteryx gracilis*, as Simpson (1971) proposed, should be considered valid but is also a *nomen dubium* according to Art. 13.1 (recommendation 13A) of the ICZN (1999) (Acosta Hospitaleche and Reguero 2011a).

During the last years, many phylogenetic proposals have been published (Clarke et al. 2003, 2007; Ksepka and Clarke 2010; Ksepka et al. 2012 and literature therein). For Sphenisciformes Clarke et al. (2003) proposed phylogenetic definitions for higher taxa within the penguin total group. Pansphenisciformes is applied to the clade including all taxa more closely related to Spheniscidae than any other extant avian lineage. Sphenisciformes is applied in a more exclusive sense to the clade including all Pansphenisciformes that share the apomorphic loss of aerial flight. Spheniscidae is applied to the crown clade of penguins, comprising the most recent common ancestor of all living penguin species and its descendants.

A more recent phylogenetic analysis of Sphenisciformes, combining morphological and molecular data, places the Weddellian penguin distantly and basally related to the extant penguin radiation (crown clade: Spheniscidae) (Ksepka and Clarke 2010). The Paleocene *Crossvallia unienwillia* remains the oldest stem record of the Weddellian sphenisciforms and provides a calibration for the basal sphenisciforms divergence at 55–56 Ma (Thanetian). The earliest record of penguins in La Meseta Formation is in the Ypresian Acantilados Allomember, deposited at 52.4 and 54.3 Ma based on $^{87/86}$Sr dates (Reguero et al. 2002). Although late Paleocene–middle Eocene sampling of the penguin record used is not complete, stratigraphically calibrated, specific-level phylogenies (i.e., Ksepka and Clarke 2010; Clarke et al. 2010) indicate that an important radiation took place during this interval in Antarctica; at minimum, five clades of sphenisciforms diverged by the early Eocene (Ksepka and Clarke 2010). The first significant radiation of the Weddellian sphenisciforms took place in the Ypresian/Lutetian

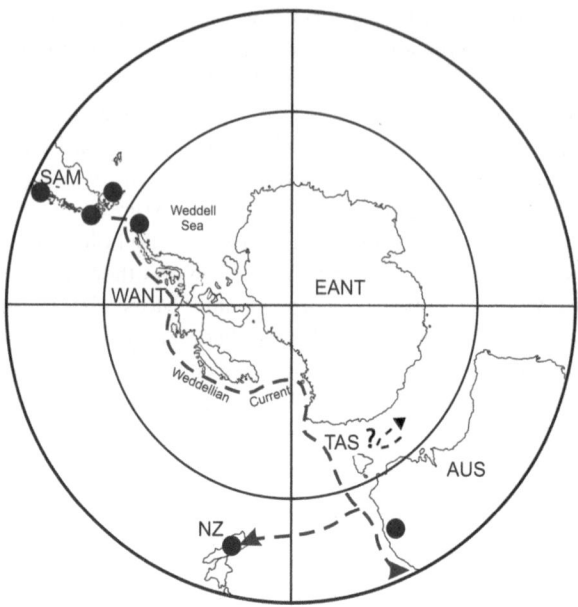

Fig. 4.7 Paleogeographic reconstruction of the southern continents (Gondwana) and the Weddellian Province at 50 Ma (Ypresian) showing the probable dispersal of Weddellian Sphenisciformes (*dotted lines* and *arrows*). *Black circles* represent fossil localities discussed in the text

Cucullaea I Allomember of the La Meseta Formation, Seymour Island with 8 sympatric species (Acosta Hospitaleche and Reguero 2011b). This Early Eocene diversity of Weddellian sphenisciformes supports several separate dispersals of giant penguins from Antarctica to lower latitudes: Australia (late Eocene, paleolatitude ~33°S, Jenkins 1974), New Zealand (late Eocene/early Oligocene, paleolatitude ~45°S, Simpson 1971), Argentina (middle Eocene, paleolatitude ~54°S, Clarke et al. 2003), Chile (middle to late Eocene, paleolatitude ~52°S, Sallaberry et al. 2010), and Peru (middle to late Eocene, paleolatitude ~14°S, Clarke et al. 2010) regions during greenhouse earth conditions (Fig. 4.7).

The highest diversity of Weddellian penguins is documented in the *Anthropornis nordenskjoeldi* Biozone (Tambussi et al. 2006) within the Priabonian Submeseta Allomember (at DPV 13/84 locality, Fig. 4.8); this unit was deposited at 34.2 Ma based on $^{87/86}$Sr dates. Sphenisciforms of the *Anthropornis nordenskjoeldi* Biozone provide the major taxonomic and body size diversity with 10–14 species co-occurring sympatrically. Discrepancies in the number of recognized species are fundamentally based on the recognition of some species such as *Archaeospheniscus lopdelli*, *Palaeeudyptes antarctica* and the two species of *Tonniornis* in the *Anthropornis nordenskjoeldi* Biozone.

Palaeeudyptes is currently considered a key taxon with biogeographic significance, is the most widespread Weddellian penguin genus in the Southern Hemisphere during the Eocene and Oligocene (Acosta Hospitaleche and Reguero 2011b, see also Ksepka et al. 2012). The FAD (First Appearance Datum) of *Palaeeudyptes gunnari* is located within the Ypresian Campamento Allomember (La Meseta Formation).

Fig. 4.8 Landscape of the northern end of Seymour Island showing the continuous surface exposure of the Submeseta Allomember (*Anthropornis nordenskjoeldi* Biozone) at DPV 13/84 locality (looking east at Weddell Sea)

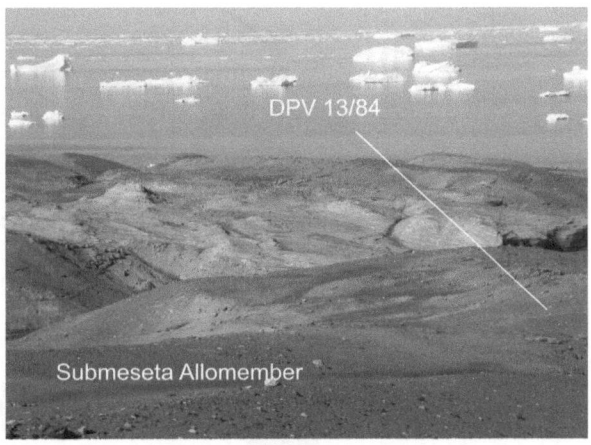

Palaeeudyptes is widely represented in West Antarctica, South America (Chile and probably Peru), New Zealand, and Australia, and is the most numerous in Argentine and Polish collections (Myrcha et al. 2002). This might be the reason for *Palaeeudyptes* having been repeatedly chosen by palaeoartists to represent the penguin fauna of Antarctica during Eocene times (see for example Mikołuszko 2007; Lloyd Spencer Davis 'Penguins' 2007; Jadwiszczak 2009). The stratigraphic distribution of this species within La Meseta Formation includes numerous remains coming from the Submeseta Allomember and a few others from the Middle Eocene *Cucullaea* I Allomember. *Palaeeudyptes gunnari* was a medium sized penguin, living exclusively in cool temperate settings during the Early and Late Eocene in Seymour Island. The highest levels (Priabonian, late Eocene, Submeseta Allomembers, ~34–36 Ma) document a major taxonomic and body size diversity of Weddellian sphenisciforms with 14 species co-occurring sympatrically.

Recently, the presence of *Palaeeudyptes* was reported in the Eocene of Chile (Sallaberry et al. 2010). The assignment of these remains was only possible after the description of the first articulated skeleton of *Palaeeudyptes* (Acosta Hospitaleche and Reguero 2010b) (Fig. 4.9). In this regard, recent findings in Peru (Clarke et al. 2010) suggest that *Palaeeudyptes*, or some other closely related taxon, would have reached the coast of Peru during the Middle Eocene.

Additionally, the discovery of a single partial skeleton of an undetermined Pansphenisciformes from the Late Eocene of Leticia Formation, Tierra del Fuego (Clarke et al. 2003) has important palaeobiogeographic significance. Based on its marine invertebrates Olivero and Malumián (1999) regarded this unit equivalent to the upper part of the La Meseta Formation.

The Eocene/Oligocene transition boundary (EOB) was characterized by sharp climatic deterioration linked to the progressive separation of South America and Antarctica and the strengthening of the Circum-Antarctic Current (Kennett 1977). The effect of these changes on the Antarctic marine fauna is unknown, but the

Crossvallia unienwillia

Palaeeudyptes gunnari

Anthropornis nordenskjoeldi

Fig. 4.9 Reconstruction of three Weddellian penguins standing, with preserved elements from MLP (Museo de La Plata) specimens: *Crossvallia unienwillia* (Cross Valley Fm., Valle de las Focas Allomember, Late Paleocene, Seymour Island, Antarctic Peninsula, Tambussi et al. 2005); *Palaeeudyptes gunnari* (La Meseta Fm., Campamento Allomember, Early Eocene, Seymour Island, Antarctic Peninsula, Acosta Hospitaleche and Reguero 2010a, b) and *Anthropornis nordenskjoeldi* (La Meseta Fm., Submeseta Allomember, Late Eocene, Seymour Island, Antarctic Peninsula, Wiman 1905a). Reconstructions and silhouettes by Pablo Motta

disappearance (extinction?) of *P. gunnari*, together with several other Eocene penguins, i.e., *Anthropornis nordenskjoeldi*, broadly coincides with these events.

The whole evolutionary history of penguins was marked by events of radiation and extinction during the Paleogene in Antarctica which explain modern penguin diversity. Stratigraphically calibrated, specific-level phylogenies of Sphenisciformes have recently been published by Ksepka and Clarke (2010) and Clarke et al. (2010). These analyses provide a simple and straightforward directional explanation for the distribution of sphenisciformes through time,

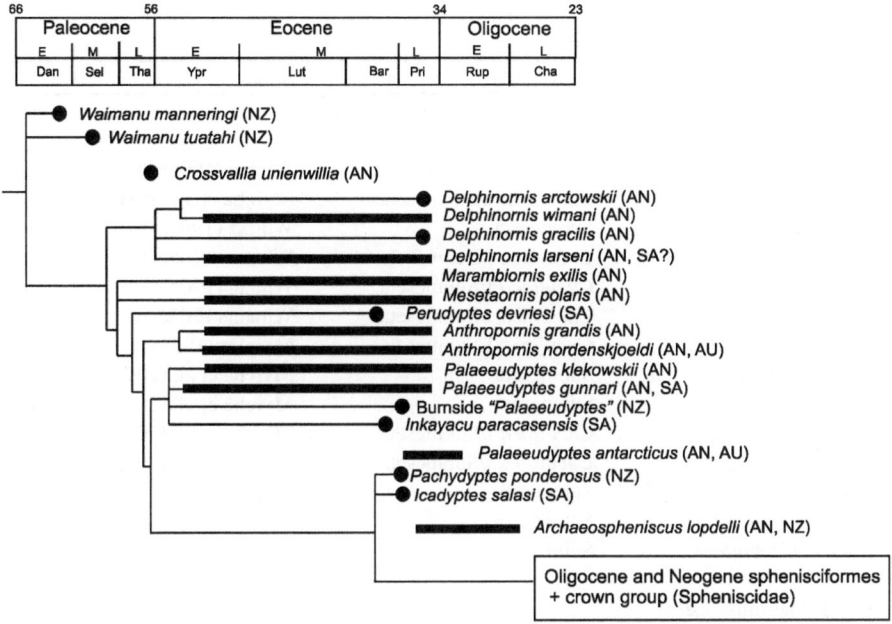

Fig. 4.10 Phylogeny of basal Spheniciformes calibrated to the Paleogene stratigraphic record, Seymour Island, Antarctic Peninsula. Phylogeny redrawn and modified from Ksepka and Clarke (2010) and Clarke et al. (2010). Most penguins (14 species) from the La Meseta Formation, Seymour Island occur in Submeseta Allomember (Telm 7 of Sadler 1988, 34.2–36.1 Ma), but a few species (8) have been reported from lower units (see Myrcha et al. 2002; Jadwiszczak 2006a), accounting for the extended ranges of *Delphinornis larseni*, *Anthropornis grandis*, and *Palaeeudyptes gunnari*. Minimum estimated times that these taxa diverged from their older sister taxa are based on ghost lineages indicated by black horizontal lines. Geographical locality for fossil taxa is provided in parentheses following taxa names. Abbreviations: *AN*, Antarctica; *AU*, Australia; *NZ*, New Zealand and surrounding islands; *SA*, South America

and are also highly consistent with stratigraphic ordering of the taxa (Fig. 4.10). However, incompleteness of the fossil record reflected in these analyses means that minimum divergence times must be established through the calculation of 'ghost lineages'.

These phylogenies reveal that important radiations of small and large Weddellian penguins took place during the Late Paleocene/Early Eocene in the Antarctic Peninsula. The Late Paleocene *Crossvallia unienwillia*, not included in the phylogenetic analyses of Ksepka and Clarke (2010) and Clarke et al. (2010), seems to have a significant role in the evolution of the Weddellian penguins. Derived features present in *Crossvallia* and all more crownward penguins, but absent in the Paleocene *Waimanu* spp. from New Zealand, i.e., a more flattened humerus and development of a trochlea for the tendon of *M. scapulotriceps* at the distal end of the humerus, indicate a significant phylogenetic connection between these species. However, the Paleocene *Waimanu* spp. from New Zealand are similarly placed as

Table 4.5 Taxonomic list, stratigraphy, and references for birds from the La Meseta Formation of Seymour Island, Antarctic Peninsula

Taxon	Stratigraphy (Allomember)	Source
Aves		
Sphenisciformes		
Palaeeudyptes gunnari	Campamento/Submeseta	Wiman (1905a, b)
Palaeeudyptes klekowskii	*Cucullaea* I/Submeseta	Myrcha et al. (1990)
Palaeeudyptes antarcticus	Submeseta	Wiman (1905a, b)
Anthropornis nordenskjoeldi	*Cucullaea* I/Submeseta	Wiman (1905a, b)
Anthropornis grandis	*Cucullaea* I/Submeseta	Wiman (1905a, b)
Delphinornis larseni	*Cucullaea* I/Submeseta	Wiman (1905a, b)
Delphinornis wimani	*Cucullaea* I/Submeseta	Marples (1953)
Delphinornis arctowskii	Submeseta	Myrcha et al. (2002)
Delphinornis gracilis	Submeseta	Myrcha et al. (2002)
Mesetaornis polaris	*Cucullaea* I/Submeseta	Myrcha et al. (2002)
Marambiornis exilis	*Cucullaea* I/Submeseta	Myrcha et al. (2002)
Archaeospheniscus lopdelli	Submeseta	Myrcha et al. (2002)
Tonniornis mesetaensis	Submeseta	Tambussi et al. (2006)
Tonniornis minimun	Submeseta	Tambussi et al. (2006)
Pelecaniformes		
Fam. Pelagornithidae		
gen. et sp. indet.	*Cucullaea* I; Submeseta	Tonni (1980)
Procellariiformes		
Fam. Diomedeidae		
gen. et sp. indet.	*Cucullaea* I	Tambussi and Tonni (1988)
Ratitae		
gen. et sp. indet.	Submeseta	Tambussi et al. (1994)
Gruiformes		
Fam. Phorusrhacidae		
gen. et sp. indet.	Submeseta	Case et al. (1987)
Falconiformes		
Fam. Falconidae		
gen. et sp. indet	*Cucullaea* I	Tambussi et al. (1995)

stem sphenisciformes by Ksepka and Clarke (2010), indicating that these taxa might also be relevant to the origin of the sphenisciform clades. The most recent common ancestor of *Crossvallia* and living penguins (Spheniscidae) is thus inferred to be present by the late Paleocene in West Antarctica (James Ross Basin).

Based on previous studies (Tambussi and Acosta Hospitaleche 2007), we recognize fifteen species of Antarctic penguins (Table 4.5). The distribution of three of them includes areas outside the James Ross Basin: *A. nordenskjoeldi* is known in sediments of the Late Eocene from Australia (Jenkins 1974; Fordyce and Jones 1990), *Palaeeudyptes antarcticus* is recorded in the Early Oligocene from New Zealand and Australia (Simpson 1971b), and *Archaeospheniscus lopdelli* is recorded in the early Late Oligocene from New Zealand (Marples 1952). Until publication of the results of Tambussi et al. (2006) describing the humeri

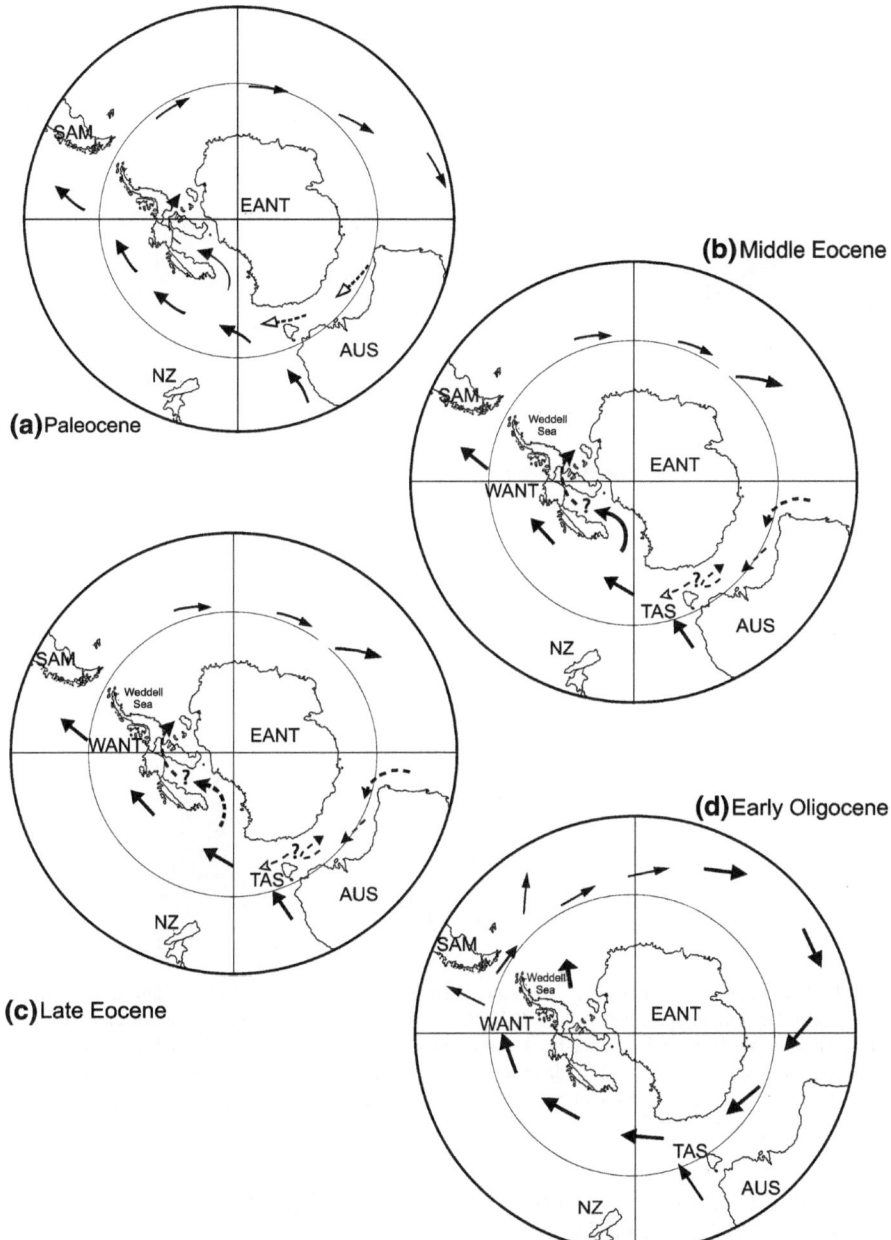

Fig. 4.11 Polar stereographic projection to 45°S of the southern ocean showing the Antarctic circulation in: **a** Paleocene, **b** Middle Eocene, **c** Late Eocene, and **e** Early Oligocene

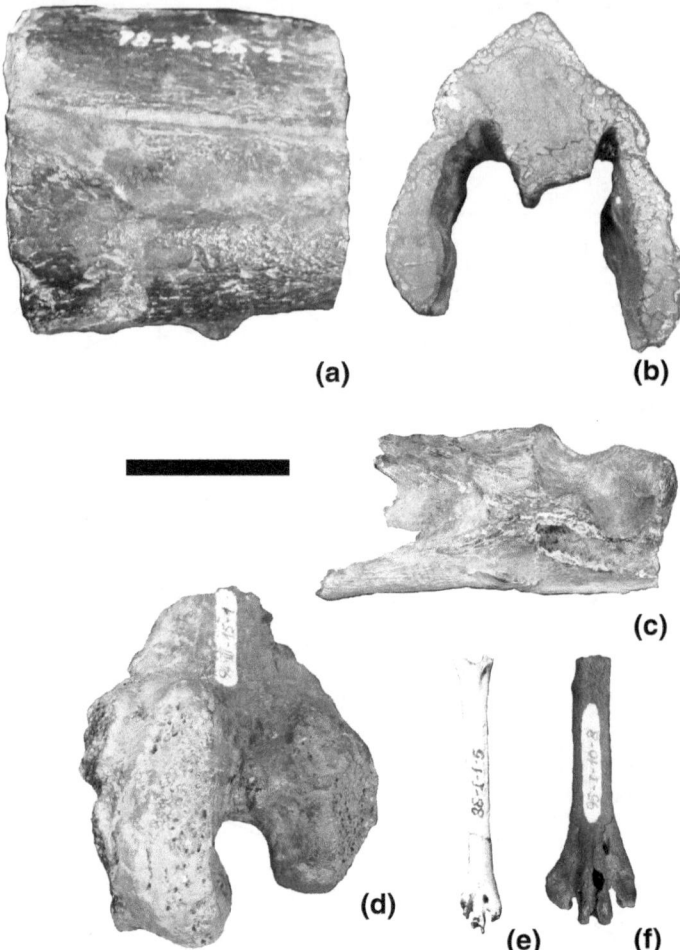

Fig. 4.12 Eocene birds from La Meseta Formation, Seymour Island, Antarctic Peninsula. Pelagornithidae, gen et sp. indet., MLP 78-X-26-1, fragment of the rostrum: **a** lateral view; **b** transversal view; **c** MLP 83-V-30-1, posterior fragment of the right mandible, lingual view; **d** Ratitae, gen. et sp. indet., MLP 94-III-15-1, distal end of right tarsometatarsus; **e** Diomedeidae, gen. et sp. indet., MLP 86-I-1-5, left tarsometatarsus; **f** Falconiformes Polyborinae, gen. et sp. indet., MLP 95-I-10-8, distal fragment of left tarsometatarsus. Scale bar = 2 cm

of Antarctic penguins, *Palaeeudyptes antarcticus* and *Archaeospheniscus lopdelli* were thought to be restricted to New Zealand and Australia.

Based on collections from Seymour Island housed in the Museo La Plata (MLP) and the Institute of Biology, University of Białystok, Poland (IB/P/B) the systematics of the Weddellian penguins are depicted in Appendix.

The Eocene/Oligocene transition was characterized by sharp climatic deterioration linked to the progressive separation of South America and Antarctica and the strengthening of the Circum-Antarctic Current (Kennett 1977). Models of ocean circulation help explain the distribution of Paleogene marine vertebrate assemblages throughout the mid to high southern palaeolatitudes, and have implications for the refinement of Antarctic palaeogeography at this time (Fig. 4.11).

The rest of the Paleogene birds that are recorded in La Meseta Formation are scarce (see Table 4.5), restricted to two cursorial birds: a ratite (Tambussi et al. 1994, Fig. 4.12d) and a phorusrhacid terror bird (Case et al. 1987). There are in addition a falconid polyborine (Tambussi et al. 2005, Fig. 4.12f), a procellariiform diomedeid (Tambussi and Tonni 1988, Fig. 4.12e) and a pelagornithid (Tonni and Tambussi 1985, Fig. 4.12a–c). The last two are flying birds whose life histories are linked mainly to coastal waters.

References

Acosta Hospitaleche C, Tambussi C, Donato M, Cozzuol M (2007) A new Miocene penguin from Patagonia and its phylogenetic relationships. Acta Palaeontologica Polonica 52:299–314

Acosta Hospitaleche C, Reguero M (2010a) Main pathways in the evolution of Antarctic fossil penguins (Seymour/Marambio Island, La Meseta Formation, Eocene): cooling events and marine circulation. In: XXXI SCAR, Open Science Conference, Buenos Aires, Argentina

Acosta Hospitaleche C, Reguero MA (2010b) First articulated skeleton of *Palaeeudyptes gunnari* from the late Eocene of Seymour Island (=Marambio) (Antarctica). Antarct Sci 22:189–298

Acosta Hospitaleche C, Reguero M (2011a) Taxonomic notes about *Ichtyopteryx gracilis* Wiman, 1905 and *Orthopteryx gigas* Wiman, 1905 (Aves, Spheniscidae). *Alcheringa* 35. doi:10.1080/03115518.2011.527476

Acosta Hospitaleche C, Reguero M (2011b) Evolution and biogeography of Paleogene Weddellian penguins (Aves: Sphenisciformes) of the James Ross Basin, Antarctic Peninsula. In: 11th international symposium on Antarctic earth sciences (ISAES), Edinburgh

Albright LB, Woodburne MO, Case JA, Chaney DS (2003) A leatherback sea turtle from the eocene of Antarctica: implications for the antiquity of gigantothermy in Dermochelyidae. J Vertebr Paleontol 23:945–949

Bargo MS, Reguero MA (1998) Annotated catalogue of the fossil vertebrates from Antarctica housed in the Museo de La Plata. I. Birds and land mammals from La Meseta formation (Eocene-? early Oligocene). In: Casadio S (ed) Paleógeno de América del Sur y de la Península Antártica, vol 5. Asociación Paleontológica Argentina, Publicación Especial, pp 211–221

Bona P, de la Fuente MS, Reguero MA (2010) New fossil turtle remains from the eocene of the Antarctic Peninsula. Antarct Sci 22:531–532

Borsuk-Bialynicka M (1988) New remains of archaeoceti from paleogene of Antarctica. Pol Polar Res 9:437–445

Brodkorb P (1963) Catalogue of fossil birds. 1 (Archaeopterygiformes through Ardeiformes). Bull Fla State Mus (Biol Sci) 7:177–293

Case JA (1988) Paleogene floras from Seymour Island, Antarctic Peninsula. In: Feldmann RM, Woodburne MO (eds) Geology and paleontology of Seymour Island, Antarctic Peninsula, vol 169. Memoir of the Geological Society of America, Antarctic, pp 523–530

Case JA (1992) Evidence from fossil vertebrates for a rich Eocene Antarctic marine environment. In: Kennett JP, Warnke DA (eds) The Antarctic paleoenvironment: a perspective on global change, vol 56. Antarctic Research Series, pp 119–130

Case JA, Martin JE, Chaney DS, Reguero M, Marenssi SA, Santillana SM, Woodburne MO (2000) The first duck-billed dinosaur (Hadrosauridae) from Antarctica. Journal of Vertebrate Paleontology 20:612–614

Case JA, Martin JE, Reguero MA (2007) A dromaeosaur from the Maastrichtian of James Ross Island and the Late Cretaceous Antarctic dinosaur fauna. U.S. Geological Survey and The National Academies; USGS OF-2007-1047, Short Research Paper, 083. doi:10.3133/of2007-1047.srp083

Case JA, Woodburne MO, Chaney D (1987) A gigantic phororhacoid(?) bird from Antarctica. Journal of Paleontology 16:1280–1284

Chatterjee S, Small BJ (1989) New plesiosaurs from the Upper Cretaceous of Antarctica. In: Crame JA (ed) Origin and evolution of the Antarctic biota, vol 47. Geological Society, London, Special Publication, pp 197–215

Cione AL, Medina F (1987) A record of *Notidanodon pectinatus* (Chondrichtyes, Hexanchiformes) in the upper cretaceous of Antarctic Peninsula. Mesoz Res 1:79–88

Cione AL, Reguero MA (1994) New records of the sharks *Isurus* and *Hexanchus* from the eocene of Seymour Island, Antarctica. Proc Geol Assoc 105:1–14

Cione AL, Reguero MA (1998) An eocene basking shark (Lamniformes, Cetorhinidae) from Antarctica. Antarct Sci 10:83–88

Cione AL, del Valle RA, Rinaldi CA, Tonni EP (1977) Nota preliminar sobre los pingüinos y tiburones del Terciario inferior de la isla Vicecomodoro Marambio, Antártida, vol 213. Contribuciones del Instituto Antártico Argentino, pp 1–21

Cione AL, Azpelicueta MM, Bellwood DR (1994) An oplegnathid fish from the Eocene of Antarctica. Palaeontology 37:931–940

Cione, AL, Reguero MA, Elliot DH (2001) A large osteichthyan vertebra from the Eocene of Antarctica. N jb Geol Paläont Mh 9:543–552

Cione AL, Reguero MA, Acosta Hospitaleche C (2007) Did the continent and sea have different temperatures in the northern Antarctic Peninsula during the middle eocene? Revista de la Asociación Geológica Argentina 62:586–596

Clarke JA, Olivero EB, Puerta P (2003) Description of the earliest fossil penguin from South America and first paleogene vertebrate locality of Tierra del Fuego, Argentina. Am Mus Novitates 3423:18

Clarke JA, Ksepka DT, Stucchi M, Urbina M, Giannini N, Bertelli S, Narváez Y, Boyd CA (2007) Paleogene equatorial penguins challenge the proposed relationship between biogeography, diversity, and Cenozoic climate change. Proc Natl Acad Sci 104:11545–11550

Clarke JA, Ksepka D, Salas-Gismondi R, Altamirano A, Shawkey MD, D'Alba L, Vinther J, DeVries TJ, Baby P (2010) Fossil evidence for evolution of the shape and color of penguin feathers. Science 330:954–957

Coria RA, Moly JJ, Reguero M, Santillana S (2008) Nuevos restos de Ornithopoda (Dinosauria, Ornithischia) de la Fm. Santa Marta, Isla James Ross, Antártida. Ameghiniana 45(Supl):25R

Cozzuol MA (1988) Comentarios sobre los Archaeoceti (Mammalia, Cetacea) de la isla Vicecomodoro Marambio, Antártida. In: Quiroga JC, Cione AL (eds) 5th Jornadas Argentinas de Paleontología Vertebrados, Abstracts, La Plata, vol 32

de la Fuente MS, Santillana SN, Marenssi SA (1995) An Eocene leatherback turtle (Cryptodira: Dermochelyidae) from Seymour Island, Antarctica. Studia Geologica Salmanticensia 31:17–30

de la Fuente M, Novas FE, Isasi MP, Lirio JM, Nuñez HJ (2010) First cretaceous turtle from Antarctica. J Vertebr Paleontol 30:1275–1278

del Valle R, Medina F, Brandoni Z (1977) Nota preliminar sobre los hallazgos de reptiles fósiles marinos del suborden Plesiosauria en las islas James Ross y Vega, Antártida, vol 212. Contribuciones del Instituto Antártico Argentino, pp 1–13

Doktor M, Gazdzicki A, Jermanska A, Porebski SJ, Zastawniak E (1996) A plant-and-fish assemblage from the Eocene La Meseta Formation of Seymour Island (Antarctic Peninsula) and its environmental implications. Palaeontol Pol 55:127–146

Fordyce RE, Jones CM (1990) Penguin history and new fossil material from New Zealand. In: Davis LS, Darby JT (eds) Penguin biology. Academic Press, San Diego, pp 419–446

Fordyce RE, Barnes LG (1994) The evolutionary history of whales and dolphins. Annual Review of Earth and Planetary Sciences 22:419–455

Fostowicz-Frelik L (2003) An enigmatic whale tooth from the upper eocene of Seymour Island, Antarctica. Pol Polar Res 24:13–28

Fostowicz-Frelic L, Gazdzicki A (2001) Anatomy and histology of plesiosaur bones from the late cretaceous of Seymour Island, Antarctic Peninsula. Palaeontologia Polonica 60:7–32

Gasparini Z, del Valle R (1981) Mosasaurios: primer hallazgo en el Continente Antártico. Antártida 11:16–20

Gasparini Z, del Valle R, Goñi R (1984) Un elasmosáurido (Reptilia, Plesiosauria) del Cretácico Superior de la Antártida, vol 305. Contribuciones del Instituto Antártico Argentino, pp 1–24

Gasparini Z, Bardet N, Martin JE, Fernández M (2003) The elasmosaurid plesiosaur *Aristonectes* cabrera from the latest Cretaceous of South America and Antarctica. J Vertebr Paleontol 23:104–115

Grande L, Chatterjee S (1987) New Cretaceous fish fossils from Seymour Island, Antarctic Peninsula. Palaeontology 30:829–837

Grande L, Eastman J (1986) A review of Antarctic ichthyofaunas in the light of new fossil discoveries. Palaeontology 29:113–137

Griffin M (1991) Eocene bivalves from the Río Turbio formation, southwestern Patagonia (Argentina). J Paleontol 65(1):119–146

Hooker JJ, Milner AC, Sequeira SEK (1991) An ornithopod dinosaur from the Late Cretaceous of West Antarctica. Antarct Sci 3:331–332

ICZN International Code of Zoological Nomenclature (1999) International trust for zoological nomenclature. The Natural History Museum, London

Jadwiszczak P (2009) Review penguin past: the current state of knowledge. Pol Polar Res 30:3–28

Jadwiszczak P, Mörs T (2011) Aspects of diversity in early Antarctic penguins. Acta Palaeontol Pol 56:269–277

Jenkins RJF (1974) A new giant penguin from the Eocene of Australia. Palaeontology 17:291–310

Jerzmanska A (1991) First articulated teleost fish from the Paleogene of West Antarctica. Antarct Sci 3:309–316

Jerzmanska A, Swidnicki J (1992) Gadiform remains from the La Meseta Formation (Eocene) of Seymour Island, West Antarctica. Pol Polar Res 13:241–253

Kellner AWA, Rodrigues Simões T, Riff D, Grillo O, Romano P, de Paula H, Ramos R, Carvalho M, Sayão J, Oliveira G, Rodrigues T (2011) The oldest plesiosaur (Reptilia, Sauropterygia) from Antarctica. Polar Research 30:1–6

Kennett JP (1977) Cenozoic evolution of Antarctic glaciation, the circum-Antarctic oceans and their impact on global paleoceanography. Journal of Geophysic Research 82:3843–3859

Kriwet J (2003) First record of Early Cretaceous shark (Chondrichthyes, Neoselachii) from Antarctica. Antarctic Science 15:519–523

Kriwet J (2005) Additions to the eocene Selachian fauna of Antarctica with comments on Antarctic selachian diversity. J Vertebr Paleontol 25(1):1–7

Kriwet K, Lirio JM, Nuñez HJ, Puceat E, Lécuyer C (2006) Late cretaceous Antarctic fish diversity. In: Francis JE, Pirrie D, Crame JA (eds) Cretaceous-tertiary high-latitude palaeoenvironments, James Ross Basin, Antarctica, vol 258. Geological Society of London, Special Publications, pp 83–100

Ksepka DT, Clarke JA (2010) The basal penguin (Aves: Sphenisciformes) *Perudyptes devriesi* and a phylogenetic evaluation of the penguin fossil record. Bulletin of the American Museum of Natural History 337:1–77

Ksepka DT, Bertelli S, Giannini NP (2006) The phylogeny of the living and fossil Sphenisciformes (penguins). Cladistics 22:412–441

Ksepka DT, Fordyce RE, Ando T, Jones CM (2012) New fossil penguins (Aves, Sphenisciformes) from the oligocene of New Zealand reveal the skeletal plan of stem penguins. J Vertebr Paleontol 32(2):235–254

Lloyd Spencer Davis 'Penguins' (2007) Te Ara the Encyclopedia of New Zealand. http://www.TeAra.govt.nz/EarthSeaAndSky/BirdsOfSeaAndShore/Penguins/en

Long DJ (1991) Fossil cutlassfish (Perciformes: Trichiuridae) teeth from the La Meseta Formation (Eocene), Seymour Island, Antarctic Peninsula. PaleoBios 13:3–6

Long DJ (1992a) The shark fauna from La Meseta Formation (Eocene), Seymour Island, Antarctic Peninsula. J Vertebr Paleontol 12:1–32

Long DJ (1992b) An Eocene wrasse (Perciformes; Labridae) from Seymour Island. Antarct Sci 4:235–237

Long DJ (1992c) Paleoecology of eocene Antarctic sharks. In: Kennett JP, Warnke D (eds) The Antarctic paleoenvironment: a perspective on global change, vol 56. Antarctic Research Series, pp 131–139

Long DJ, Stilwell JD (2000) Fish remains from the eocene of mount discovery, East Antarctica, vol 76. Antarctic Research Series, pp 349–353

Marenssi SA, Santillana SN, Rinaldi CA (1998a) Paleoambientes sedimentarios de la Aloformación La Meseta (Eoceno), Isla Marambio (Seymour), Antártida, vol 464. Instituto Antártico Argentino, Contribución, p 51

Marenssi SA, Santillana SN, Bauer M (2012) Estratigrafía, petrografía sedimentaria y procedencia de las formaciones Sobral y Cross Valley (Paleoceno), isla Marambio (Seymour), Antártida. Andean Geology 39(1):67–91

Marples BJ (1952) Early tertiary penguins of New Zealand. New Zealand Geological Survey. Palaeontol Bull 20:1–66

Marples BJ (1953) Fossil penguins from the mid-Tertiary of Seymour Island. Falkland Islands Dependencies Survey Scientific Reports 5:1–15

Martin JE (2006) Biostratigraphy of the Mosasauridae (Reptilia) from the Cretaceous of Antarctica. In: Francis JE, Pirrie D, Crame JA (eds) Cretaceous-tertiary high-latitude palaeonvironments, James Ross Basin, Antarctica, vol 258. Geological Society of London, Special Publications, pp 101–108

Martin JE, Crame JA (2006) Palaeobiological significance of high-latitude Late Cretaceous vertebrate fossils from James Ross Basin. Antarctica. In: Francis JE, Pirrie D, Crame JA (eds) Cretaceous-tertiary high-latitude palaeonvironments, James Ross Basin, Antarctica, vol 258. Geological Society of London, Special Publications, pp 109–124

Martin JE, Bell G Jr, Case JA, Chaney DS, Fernández MS, Gasparini Z, Reguero MA, Woodburne MO (2002) Late cretaceous mosasaurs (Reptilia) from the Antarctic Peninsula. In: Gamble JA, Skinner DNB, Henrys S (eds) Antarctica and the close of the millenium, 8th international symposium on Antarctic earth sciences, Bulletin of the Royal Society of New Zealand Bulletin vol 35, pp 293–299

Martin JE, Sawyer JF, Reguero M, Case JA (2007) Occurrence of a young elasmosaurid plesiosaur skeleton from the Late Cretaceous (Maastrichtian) of Antarctica. In: Cooper, AK, Raymond, CR (eds) Online proceedings of the 10th international symposium on Antarctic earth sciences. Antarctica: a keystone in a changing world. USGS Open-File Report 2007-1047. Short Research Paper 066. doi:10.3133/of2007-1047.srp066

Mikołuszko W (text; key illustration [a reconstruction of *Palaeeudyptes klekowskii*] by D. Cyranowska, photo and scientific consultation by P. Jadwiszczak) (2007) Wyspa pingwinów, vol 8(95). National Geographic Polska

Mitchell ED (1989) A new cetacean from the late Eocene La Meseta Formation, Seymour Island, Antarctic Peninsula. Can J Fish Aquat Sci 46:2219–2235

Molnar RE, Angriman LA, Gasparini Z (1996) An Antarctic cretaceous theropod. Mem Qld Mus 39:669–674

Myrcha A, Tatur A, del Valle R (1990) A new species of fossil penguin from Seymour Island, West Antarctica. Alcheringa 14:195–205

Myrcha A, Jadwiszczak P, Tambussi C, Noriega J, Gazdzicki A, Tatur A, del Valle RA (2002) Taxonomic revision of Eocene Antarctic penguins based on tarsometatarsal morphology. Pol Polar Res 23:5–46

Novas FE, Cambiasso AV, Lirio JM, Núñez HJ (2002) Paleobiogeografía de los dinosaurios polares de Gondwana. Ameghiniana 39 (Supl):15R

Olivero EB, Malumián N (1999) Eocene stratigraphy of southeastern Tierra del Fuego, Argentina. Am Assoc Pet Geol Bull 83(2):295–313

Olivero EB, Ponce JJ, Marsicano CA, Martinioni DR (2007) Depositional settings of the basal López de Bertodano Formation, Maastrichtian, Antarctica. Revista de la Asociación Geológica Argentina 62:521–529

Reguero MA, Gasparini Z (2007) Late cretaceous-early tertiary marine and terrestrial vertebrates from James Ross Basin, Antarctic Peninsula: a review. In: Rabassa J, Borla ML (eds) Antarctic Peninsula and Tierra del Fuego: 100 years of Swedish-Argentine scientific cooperation at the end of the world. Taylor Francis, London, pp 55–76

Reguero MA, Marenssi SA, Santillana SN (2002) Antarctic Peninsula and Patagonia Paleogene terrestrial environments: biotic and biogeographic relationships. Palaeogeogr Palaeoclimatol Palaeoecol 179:189–210

Reguero M, Tambussi CP, Mörs T, Buono M, Marenssi SA, Santillana SN (2011) Vertebrates from the basal horizons (Ypresian to Lutetian) of the *Cucullaea* I Allomember, La Meseta Formation, Seymour (Marambio) island, Antarctica. In: 11th international symposium on Antarctic earth sciences (ISAES), Edinburgh

Rich T, Vickers-Rich P, Fernández M, Santillana S (1999) A probable hadrosaur from Seymour Island, Antarctica Peninsula. In: Tomida Y, Rich T, Vickers-Rich P (eds) Proceedings of the second Gondwana dinosaur symposium. National Science Museum, Tokyo, pp 219–222

Richter M, Ward DJ (1990) Fish remains from Santa Marta Formation (Late Cretaceous) of James Ross Island, Antarctica. Antarct Sci 2:67–76

Sadler P (1988) Geometry and stratification of uppermost Cretaceous and Paleogene units of Seymour Island, northern Antarctic Peninsula. In: Feldmann RM, Woodburne MO (eds) Geology and paleontology of Seymour Island, Antarctic Peninsula, vol 169. Geological Society of America, Memoir, pp 303–320

Salgado L, Gasparini Z (2006) Reappraisal of an ankylosaurian dinosaur from the Upper Cretaceous of James Ross Island (Antarctica). Geodiversitas 28:119–135

Sallaberry MA, Yury-Yáñez RE, Otero RA, Soto-Acuña S, Torres GT (2010) Eocene birds from the western margin of southernmost South America. J Paleontol 84(6):1061–1070

Simpson GG (1946) Fossil penguins. Bull Am Mus Nat Hist 87(1):1–99

Simpson GG (1971a) Review of fossil penguins from Seymour Island. Proc R Soc Lond B Biol Sci 178:357–387

Simpson GG (1971b) A review of the pre-Pliocene penguins of New Zealand. Bulletin of the American Museum of Natural History 144:319–378

Stahl BJ, Chatterjee S (1999) A Late Cretaceous chimaerid (Chondrichthyes, Holocephali) from Seymour Island, Antarctica. Palaeontology 42:979–989

Stahl BJ, Chatterjee S (2002) A late cretaceous callorhynchid (Chondrichthyes, Holocephali) from Seymour Island, Antarctica. J Vertebr Paleontol 22:848–850

Tambussi CP, Tonni EP (1988) Un Diomedeidae (Aves, Procellariiformes) del Eoceno tardío de la Antártida. In: Quiroga JC, Cione AL (eds) 5th Jornadas Argentinas de Paleontología Vertebrados, Abstracts, La Plata, vol 4

Tambussi C, Acosta Hospitaleche C (2007) Antarctic birds (Neornithes) during the Cretaceous-Eocene times. Revista de la Asociación Geológica 62(4):604–617

Tambussi CP, Noriega JI, Gazdzicki A, Tatur A, Reguero MA, Vizcaíno SF (1994) Ratite bird from the Paleogene La Meseta Formation, Seymour Island, Antarctica. Pol Polar Res 15:15–20

Tambussi CP, Noriega JI, Santillana SN, Marenssi SA (1995) Falconid bird from the middle Eocene La Meseta Formation, Seymour Island, West Antarctica. J Vertebr Paleontol 15(supplement to 3):55A

Tambussi CP, Reguero MA, Marenssi SA, Santillana SN (2005) The earliest known penguin and the evolution of spheniscid size. GEOBIOS 38:667–675 Wiman C (1905a) Vorfläufige Mitteilung über die alttertiaren Vertebraten der Seymourinsel. Bull Geol Inst Uppsala 6:247–25

Tambussi CP, Acosta Hospitaleche CI, Reguero MA, Marenssi SA (2006) Late eocene penguins from West Antarctica: systematics and biostratigraphy. In: Francis JE, Pirrie D, Crame JA (eds) Cretaceous-tertiary high-latitude palaeoenvironments, James Ross Basin, Antarctica, vol 258. Geological Society, Special Publications, London, pp 145–161

Tonni EP (1980) Un pseudodontornítido (Pelecaniformes, Odontopterygia) de gran tamaño, del Terciario temprano de Antártida. Ameghiniana 17:273–276

Tonni EP, Tambussi CP (1985) Nuevos restos de Odontopterygia (Aves, Pelecaniformes) del Terciario temprano de Antártida. Ameghiniana 21:121–124

Ward DJ, Grande L (1991) Chimaeroid fish remains from Seymour Island, Antarctic Peninsula. Antarct Sci 3:323–330

Welton B, Zinsmeister WJ (1980) Eocene neoselachians from the La Meseta Formation, Seymour Island, Antarctic Peninsula. Contrib Sci, Nat Hist Mus Los Angeles Ctry 329:1–10

Wiman C (1905a) Über die alttertiären Vertebraten der Seymourinsel. Wissenschaftliche Ergebnisse der Schwedischen Südpolar Expedition 1901–1903(3):1–37

Wiman C (1905b) Vorfläufige Mitteilung über die alttertiaren Vertebraten der Seymourinsel. Bull Geol Inst Uppsala 6:247–253

Woodburne MO, Zinsmeister WJ (1984) The first land mammal from Antarctica and its biogeographic implications. J Paleontol 54:913–948

Woodward AS (1908) On fossil fish-remains from Snow Hill and Seymour Islands. Wissenschaftliche Ergebnisse der Schwedischen Südpolar-Expedition 1901–1903(3):1–4

Zinsmeister WJ (1979) Biogeographic significance of the Late Mesozoic and Early Tertiary molluscan faunas of Seymour Island (Antarctic Peninsula) to the final breakup of Gondwanaland. In: Gray J, Boucot A (eds) Historical biogeography, plate tectonics and the changing environment, Proceedings, 37th annual biological colloquium and selected papers. Oregon State University Press, Corvallis, pp 349–355

Zinsmeister WJ (1982) Late cretaceous-early tertiary molluscan biogeography of southern circum-Pacific. J Paleontol 56:84–102

Chapter 5
The Terrestrial Biotic Dimension of West Antarctica

5.1 West Antarctica Paleoflora

Unlike animals, which may migrate to more favorable places to live or modify their behaviors if the environmental conditions change, plants normally cannot choose similar strategies. Alternatively the plants have mechanisms of rapid morphological and adaptive changes which make them a useful source of information on the effects of ecological constraints (so true ecotypes sensu Margalef 1983). By analogy, plant fossils may be an important tool to reconstruct the past climates. Furthermore, the floras are in general prevented in their dispersion by the oceanic barriers (with few exception like proposed by Pole 1994), and represent an important source for paleogeographical reconstructions.

The diversity and distribution of paleobotanical and palynological records not only allows us to follow the paleoclimatic and paleogeographical events and their consequences, but also provides an independent tool for testing the past climate proxies developed from marine faunas and floras (Behrensmeyer 1992). By analyzing the long-term dynamics of the vegetation in WANT since the end of Cretaceous (Birkenmajer and Zastawniak 1989; Askin 1992; Dutra 2004; Birkenmajer et al. 2005; Cantrill and Poole 2002, 2005; Francis et al. 2007; Barreda and Palazzesi 2007), it is possible to detect the onset of warm temperatures as well as climatic deterioration, regionally expressed in the Paleogene of this area. The effect of these processes led to the modern configuration and distribution of the extant subtropical and temperate biomes of the Southern Hemisphere.

Here we analyze the vegetation record through the end of Cretaceous-Paleogene in West Antarctica, based on a compilation of paleobotanical and palynological records from both sides of the Antarctic Peninsula (James Ross Basin, Weddell Sea to the east, and King George Island to the west, Fig. 5.1) and its comparison with fossil floras from Australasia and southernmost South America (Fig. 5.2). We believe that this is the key to understand of what occurred in continental areas during these times, and will help to answer, in the long-term,

M. Reguero et al., *Late Cretaceous/Paleogene West Antarctica Terrestrial Biota and Its Intercontinental Affinities*, SpringerBriefs in Earth System Sciences, DOI: 10.1007/978-94-007-5491-1_5, © The Author(s) 2013

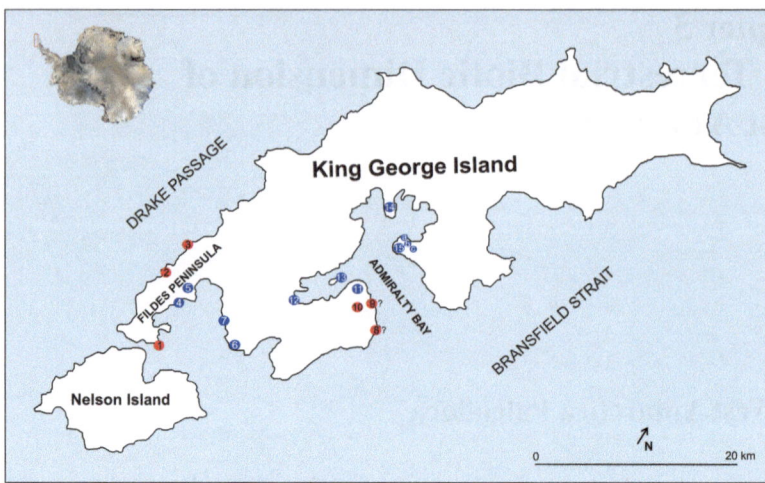

Fig. 5.1 Location map of King George Island and the localities with plant fossils. In red those with a proposed Late Cretaceous flora, in blue those corresponding to the Late Paleocene-Early Eocene, and in yellow those of Middle Eocene age. *1* Half Three Point; *2* Skua Bay (or Winckle Point); *3* Price Point; *4* Rocky Cove-Highlands; *5* Collins Glacier; *6* Potter Peninsula (Three Brothers Hill); *7* Barton Peninsula; *8* Creeping Slope; *9* Block (or Agat Point); *10* Zamek Hill; *11* Point Thomas and Petrified Forest Creek; *12* Admiralen Peak; *13* Dufayel Island; *14* Keller Peninsula; *15* Point Hennequin: **a** Dragon Glacier, **b** Smok, **c** Mount Wawel

the question postulated by Cione et al. (2007) concerning to the use of distinct data (marine and terrestrial) in past climate and paleogeographical analyses.

Nothofagus is an important paleoclimate and paleogeographical indicator (Hill and Jordan 1993; Premoli 1996; Premoli et al. 2010). This genus has been the main angiosperm represented in the paleoflora of austral regions since the Cretaceous (Fig. 5.3). Its distribution nearly corresponds to the limits of the Weddellian Province (Zinsmeister 1979; Baldoni and Askin 1993), or the South Polar Province, based on the analysis of dinoflagellate cysts (Bowman et al. 2012). The northern Antarctic Peninsula seems to be the place of origin of this genus; it is supported by putative fagalean leaf impressions found in the Lower Cretaceous of Alexander Island (Cantrill and Nichols 1996), associated to taxa that indicate the arrival of primitive angiosperms to the Antarctic Peninsula.

The origin of this genus go back to the Coniacian of Hidden Lake Formation, James Ross Island, where was identified a leaf morphotype of "unknown affinities" (Morphotype 10 of Francis et al. 2007) related to *Nothofagus*. Additional data related to the history of the *Nothofagus* lineage comes from the primitive palynomorph *Nothofagidites endurus* Stover and Evans and *N. senectus* Dettmann and Playford (Dettmann et al. 1990; Dolding 1992; Askin 1992; Keating 1992), which were reported from Late Campanian–Maastrichtian beds from the James Ross Basin and King George Island. They are associated with Late Cretaceous macrofloral assemblages containing leaves apparently related to *Nothofagus* (Zhou

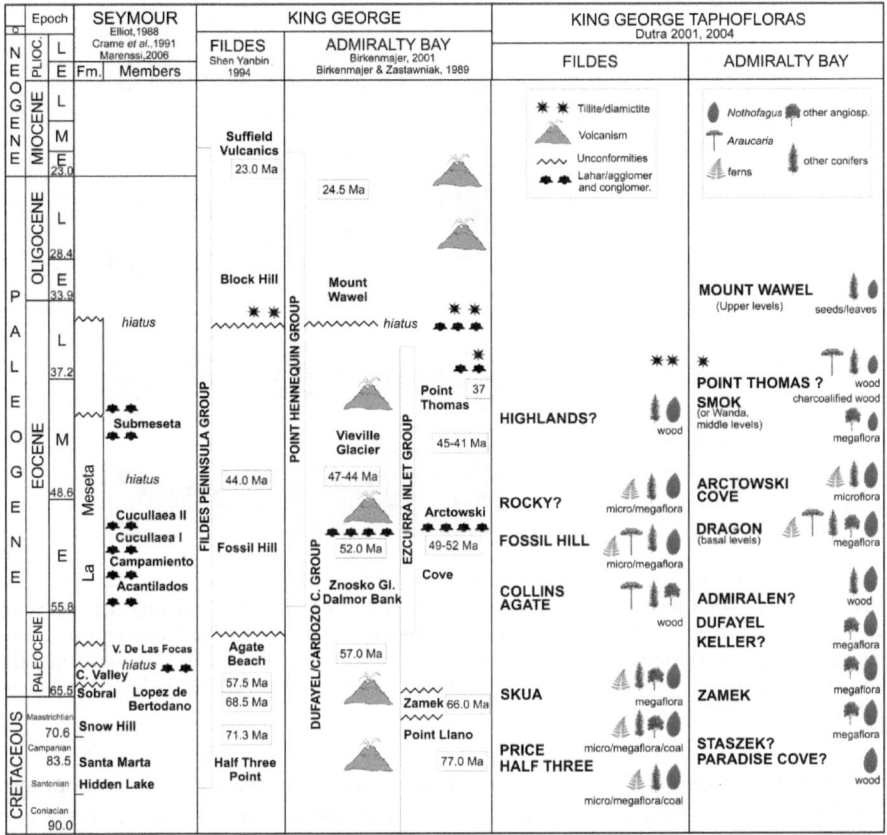

Fig. 5.2 Proposed chronology based on plant bed deposits at King George Island (Fildes Peninsula and Admiralty Bay) and its major content in plant groups (modified from Dutra 2004). Radiometric ages from Birkenmajer et al. (1986), Soliani Jr et al. (1988) and Yinxi and Yanbin (1994). Tectonically and volcanic induced deposits follow the proposition of Birkenmajer (2001). Glacial influenced lithologies (tilittes) based on Santos et al. (1990) and Birkenmajer et al. (2005). Time scale according to Gradstein et al. (2004) and Walker and Geissman (2009)

Zhiyan and Li Haomin 1994a; Dutra and Batten 2000), and thus attest to its first diversification in the WANT area (Fig. 5.4).

The increasing *Nothofagus* diversity in this region was probably stimulated by a cool climate interval at the end of Cretaceous, following the mid-Late Cretaceous warm peak attested by growth rings in fossil woods and pollen assemblages (Askin 1990; Francis and Poole 2002; Poole et al. 2005).

These macrofloras are associated to groups like Proteaceae, Cunoniaceae and Sapindaceae families of angiosperms (a condition that will return after the end of Paleogene) and to Podocarpaceae and Cupressaceae families of conifers (Fig. 5.5). When we examine the associated microfloristic record, we find that

Fig. 5.3 Ancient and modern physiognomies of the genus *Nothofagus*. **a** *Nothofagus betulifolia* Dutra 2000a, b, from the Paleocene/Eocene, Fossil Hill Formation, King George Island. Scale bar 1 cm; **b** *Nothofagus alessandrii* Espinosa, a rare and endemic form that today grows in frost free and low temperate areas of Coastal Cordillera (35°S of latitude), Chile, and considered the most primitive extant form of the genus (Copyright © 2005–2009 Michail Belov, http://www.chilebosque.cl)

to the end-Cretaceous floras share elements with both West and East Southern Hemisphere paleofloras, a condition that will be observed until the Eocene and is suggesting that an enhanced dispersion occurred in the South Polar Province. Dettmann (1989) suggested that Cretaceous Antarctic plant communities were the origin of modern austral floras.

In the spore and pollen floras of the Campanian-Paleocene of Seymour Island and King George Island, the podocarpaceous are the most abundant conifer pollen and is associated to minor angiosperms and cryptogam spores. The presence of probable Palmaceae and Liliaceae in the transitional environments of the James Ross Basin is confirming the oceanic influence (Askin 1990; Askin et al. 1991; Partridge 2002).

After nearly 10 million years the warmer conditions returned to high latitudes and environmental and floristic diversity was recaptured, as it is reflected in the Paleocene-Eocene sedimentary sequences of Seymour and King George islands (Fig. 5.6). The expansion of the tropical belts up to 38°S, with belts broader than in the Early Eocene, affected the high latitude floras that reached its maximum diversity at this time (Askin 1989; Gayó et al. 2005).

In the Late Paleocene Cross Valley Formation of Seymour Island the presence of a cool-warm temperate mixed conifer-broad-leaved evergreen and deciduous forest, dominated by large trees of *Nothofagus*, araucaria and podocarp conifers, and understorey shrubs and monocots (including a probable palm) is suggested by

Fig. 5.4 The fossil record
of *Nothofagidites* Erdtman
ex Potonié pollen grains
detaching its first occurrence,
with the primitive forms *N.
senectus* in the Santonian-
Campanian boundary
of the northwestern
Antarctic Peninsula area,
and its dispersion to other
southern Hemisphere land
masses. Abbreviations: *a
Nothofagidites* ancestral
type a (*N. senectus*); *b
Nothofagidites* ancestral
type b (*N. endurus* and
N. kaitangataensis); *B
Brassii* type (subgenus
Brassospora); *M Menziesii*
type (corresponding to
Lophozonia subgenus);
F Fusca type, in part
corresponding to *Fuscospora*
subgenus, the other to
Nothofagus subgenus (**N**)

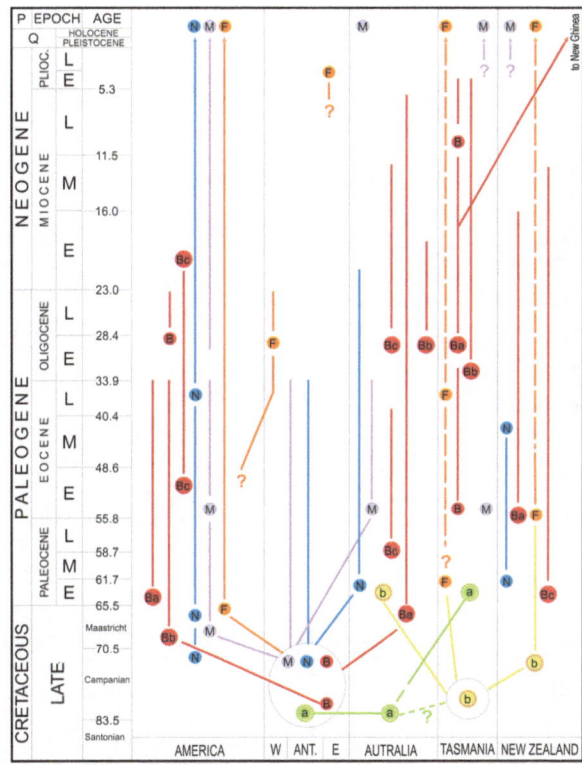

Fig. 5.5 *Coniopteris* sp.
(Dicksoniaceae), with
other pteridophyte groups,
a main component in the
Late Cretaceous deposits
of northwestern of the
Antarctic Peninsula. Scale
bar = 10 mm

Fig. 5.6 Some leaf impressions from the late paleocene-early eocene deposits from King George Island. **a** Cupressaceae; **b** Podocarpaceae; **c** Araucariaceae; **d** a 3D preserved *Araucaria* shoot with original woody spaces replaced by calcite; **e** a big leaf of *Nothofagus*; **f** laurophyllic morphotype; **g** Malvaceae (Sterculiaceae); **h** Myrtaceae; **i** Sapindaceae/Cunoniaceae

the fossil assemblages (Askin 1997; Poole et al. 2005). According to Francis et al. (2007) the Antarctic Late Paleocene floras indicate warm climates probably without ice, even in winter. In southernmost South America these sorts of mixed floras (a blend of cold and warm taxa with Austral-Antarctic and Neotropical affinities) are recorded since the beginning of the Eocene (Gayó et al. 2005), attesting a diachronic pattern in relation to Antarctic Peninsula areas. Modern analogues to these communities, which also characterize the Early Eocene paleoflora of Laguna del Hunco (Argentina) and Pichileufú (Chile), are the Valdivian Forest in Chile (Poole et al. 2001) and the *Araucaria angustifolia* woodland, the southernmost expression of Atlantic Rainforest in southern Brazil.

On the western flank of the Antarctic Peninsula, Late Paleocene and Early Eocene paleofloras were restricted to the King George Island. The ages and the distinction of these floristic assemblages in the Paleocene and Eocene are difficult to determine. K–Ar radiometric age data could be affected by processes of reheating the lithologies in the volcanic context of the island (Birkenmajer 1981; Soliani and Bonhomme 1994). The Late Paleocene age for assemblages at King George Island is suggested by a suite of age data between 57–59 Ma (Pankhurst and Smellie 1983; Shen 1994) obtained in some localities, i.e., Potter and Keller Peninsula, Admiralen Peak and Skua Bay (Winkle Point), the last also shares common plant fossils with Late Cretaceous levels in the island. The flora present in the basal part of Mount Wawel (Dragon Flora) succession, or all Fossil Hill succession, by its high diversified assemblages, large *Nothofagus* leaves and Araucariaceae-Cupressaceae-Podocarpaceae conifers, also could corresponds to this time (Birkenmajer and Zastawniak 1989; Torres 1990; Li Haomin 1994; Poole et al. 2001; Dutra 2004; Fontes and Dutra 2010).

The transition to the Early Eocene is marked by the the influx, in the Nothofagaceae, Podocarpaceae and Cunoniaceae dominated vegetation, of some subtropical to temperate angiosperms (Lauracae, Myrtaceae, Malvaceae or Sterculiaceae, Elaeocarpaceae and Anacardiaceae) and the maintainance in the macrofossils of the three conifer groups that today characterize the Southern Hemisphere floras. The Early Eocene floras also shares a great identity with coeval floras from the southernmost South America, confirming the presence of mixed mesophytic forests like proposed by Gandolfo et al. (1998a, b). And despite the fall in temperature as was suggested by those authors (Gandolfo et al. 1998a, b; Francis et al. 2007), it seems equally likely that this composition reflects either water deficiency or seasonal conditions. The increase in monocot representatives (mainly Poaceae) in the microflora, also support this interpretation (Stuchlik 1981).

Similar conditions characterize the Rio Turbio Formation in Argentina of Middle Eocene age (Berry 1937; Hünicken 1955; Archangelsky 1972; Romero and Zamaloa 1985), and the Late Eocene-Early Oligocene Sloggett Formation from Tierra del Fuego (Panti et al. 2008), confirming the affinity between the basal Eocene horizons of King George Island with the subsequent ones from South America. In Antarctic Eocene paleofloras, *Nothofagus*, when compared with older assemblages shows a reduced dominance. In the Eocene *Nothofagus* is represented by big leaves, exclusive of these Antarctic floras and with no analog in either modern or other fossil floras. These Antarctic floras are restricted to the Antarctic Peninsula and include the pioneering material studied by Dusén (1908), revised by Tanai (1986) from Seymour Island (e.g., *N. subferruginea* and *N. densinervosa*). However, there were other elements present in the Antarctic paleofloras that were preserved in several western Patagonian paleofloras (Laguna del Hunco, Pichileufú, Quinamávida and Lota-Coronel). It is interpreted as the presence of more favorable conditions for the growing of *Nothofagus* at this time in the northern Antarctic Peninsula area, and the existence a corridor of dispersion through southwest South America still functioning.

The similarity among the floristic composition and diversity of some Early Eocene South American floras (Wilf et al. 2005) and those from the Late Paleocene Cross Valley Formation, Seymour Island, suggests that there was more protected conditions close to the back-arc areas, in relation to Pacific moist winds areas, favoring the maintaining of warm and drier conditions and the grow of the mixed floras. It seems to be stimulated by the onset interval of the Paleocene–EoceneThermal Maximum (PETM) and a global negative carbon isotope excursion (Zachos et al. 2008). In the Antarctic Peninsula it also coincides with the appearance of *Araucaria* related forms in the macroflora, i.e., Araucariaceae pollen grains were ever present (but rare). Three-dimensional preservation of *Araucaria* leaves, with the original organic spaces replaced by calcite, are identical to those found by Stephens (2008) in the Campamento Allomember of the La Meseta Formation. Francis et al. (2007) suggest that this may be due to taphonomic processes, rather than a climate signal, as fragile leaves, including those of *Nothofagus*, are usually absent from the La Meseta taphoflora. On the contrary, Gandolfo et al. (1998a, b) indicated that presence of *Nothofagus* in nearly all the sequence and mainly within the *Cucullaea* I Allomember. Another important conifer element is the Podocarpaceae, recorded in some King George deposits with elements of both *Podocarpus* and *Dacrydium* s.l. (Zhian and Haomin 1994b; Birkenmajer and Zastawniak 1989; Fontes and Dutra 2010). Today this group shows a disjunct distribution. *Podocarpus*, characterized by leaf shoots with bilateral, large and openly arranged needle-like leaves, is rare in Australasia region, meanwhile the short, imbricate bifacial leaves from the *Dacrydium* s.l., grows today mainly in the "Antarctic forests" from New Zealand and Tasmania. Podocarpaceae can disperse over long distances and resist frost, guaranteed its long-term survival and wide distribution during geological time (Cubitt and Molloy 1994), but at the same time makes this group less useful for paleogeographic reconstruction. Podocarps, together with *Nothofagus*, survived close to South Pole areas (Transantarctic Mountains) until the Pliocene (Francis and Hill 1996).

Several authors, based on Seymour Island floras and using for comparison its nearest living relatives, suggested a cool-and-seasonal climate for the Early/Middle Eocene of the Antarctic Peninsula (Gandolfo et al. 1998a, b; Francis et al. 2007). Rather than taphonomic bias or climate induced conditions, the scarcity of *Nothofagus* fossils on the eastern side of the Antarctic Peninsula, could be explained as a result of the more distant location of James Ross Basin in relation to volcanic terraines (to which *Nothofagus* is preferentially adapted) or to oceanic winds. A similar situation is observed when the plant fossil assemblages from western and eastern South American basins are compared, showing that those confined to the back arc environment are poorer in *Nothofagus* and *Araucaria* remains.

Oxygen isotope analyses of marine molluscs in the same concretions that furnished the *Araucaria* shoots also yielded cool paleotemperatures of 8.3–12.5 °C, which is comparable to other estimates made for Acantilados-Campamento (Telm 3) allomembers made by Pirrie et al. (1998) and Dingle et al. (1998), or still slightly lower (Dutton et al. 2002). So, the transitional marine-continental context of Seymour Island deposits could be an excellent place to compare the responses of terrestrial and marine life to climatic changes, and could be independently used to evaluate major variables of climate parameters.

The unique sequence on the western flank of the Antarctic Peninsula that preserves a nearly complete record of Late Paleocene?-Eocene plants is Mount Wawel (or Point Hennequin, see Birkenmajer 2001), at Admiralty Bay, King George Island, that shows an impoverishment of the plant communities by the Middle Eocene (Zastawniak 1981; Zastawniak et al. 1985; Dutra 2004; Fontes and Dutra 2010). There is clear dominance of *Nothofagus* in the upper assemblages that is represented by short, perennial leaves, like those found in *N. betuloides* and *N. pumilio* (Premoli 1996; Dutra 2004). In the lower Smok plant beds (middle part of Mount Wawel sucession), *Nothofagus* is associated with representatives of Cunoniaceae/Sapindaceae (*Cupania* and *Weinmania*), Anacardiaceae, Saxifragaceae and Rosidae (*Rubus* sp., *Acaena*-like), conforming the forest understorey and reminiscent of Patagonian vegetational fossil assemblages (Fig. 5.7). At the top (upper levels of Mount Wawel) *Nothofagus* is the sole angiosperm in the assemblages, linked with some shoots and seeds of Podocarpaceae (Zastawniak et al. 1985). The reduction of leaf sizes of *Nothofagus,* in the upper levels (5–2 cm large), when compared with that in older Eocene and Paleocene levels (12–8 cm large) is of special significance. Also in evidence is the change of habit, with the appearance of perennial forms. According to Dutra (1997), most of the modern perennial *Nothofagus* points to a more rigid climate than the deciduous ones. Induced by each time more frost climates the *Nothofagus* representatives seems to change the leaf fall and more complex and delicate leaves (with many teeth and organized veins), by perennial and sclerophyllic ones (e.g., *N. solandri* and *N. truncate*, today native from New Zealand).

All those elements indicate a substantial and gradative cooling of climate to the latest Middel Eocene-Early Oligocene (?) levels of the King George Island successions, and support its close relation with Early Oligocene floras from Loreto and Río Guillermo localities in Santa Cruz, Argentina (Dusén 1907; Frenguelli 1941; Hünicken 1955; Panti 2011). The close similarities of the floristic communities from the upper levels of King George Island, and those that today grow in New Zealand are suggesting a vicariant biogeography of the genus. Pole (1994) proposed a capacity for long-distance dispersion.

Fig. 5.7 Sapindaceae related leaf (**a**) and microphyllic leaves of *Nothofagus* (**b**) from the Late Eocene-Early Oligocene(?) beds from the top of Mount Wawel succession, Admiralty Bay, King George Island. Scale bar = 1 cm

The Seymour Island deposition studied by Francis et al. (2007) also pointed to a 47 % decreasing in diversity in the Middle Eocene flora of *Cucullaea* I Allomember (La Meseta Formartion), compared to Early Eocene. Once again *Nothofagus* is the dominant form, with lesser presence of Proteaceae. The poor assemblages from the upper levels of the La Meseta Formation (Submeseta Allomember) suggest a similarity with those from King George Island and South America, and could be indicating drier climatic conditions at the end of Paleogene in eastern Patagonia produced a rapid change in vegetation (Barreda and Palazzesi 2007), preventing the invasion of wet adapted Antarctic floras in easternmost areas of Argentina. Meanwhile, easy dispersion was maintained in western areas of Andean Cordillera, mainly affecting some *Nothofagus*, Podocarpaceae, Araucariaceae, Proteaceae and terrestrial vertebrates.

Families like Myrtaceae, Malvaceae (Sterculiaceae), Cunioniaceae, Elaeocarpaceae, Fabaceae and Poaceae were components of the Antarctic Peninsula flora and they suggest warmer and drier climates. In South America and Australia the reduction of the humidity (aridization) after the Oligocene stimulated a northward migration of these groups, making them important elements in modern Southern Hemisphere landscapes (Romero 1986; Troncoso and Romero 1998; Dutra and Batten 2000; Hinojosa et al. 2006). As many previous researchers have remarked, the paleobotanical record to the Paleogene from Antarctica is strongly marked by a vegetation turnover, mainly it is documented in the loss of warm-temperate taxa and the increase of those of microphyllic character.

An Oligocene–Miocene microflora preserved at Cape Melville, King George Island, attest the survival of some sclerophyllous elements since after regional marine-based grounded ice was formed on the continental shelf, composed by gimnosperms (Cupressacea or Taxodiaceae), few pteridophyta (*Cicatricosisporites, Cyathidites*) and fungal spores, accompanied by Nothofagaceae (*Nothofagidites flemingii and Nothofagidites* Fusca group) and Chenopodiaceae (Troedson and Riding 2002). It is the last and youngest record of vegetation over the Antarctic Peninsula areas.

The analysis of these Antarctic floras confirms: (1) the reaction of plant communities against unfavorable conditions, the time involved in its dispersion to more favorable places, in relation to the information provided for the faunas, and also explain the diachronic process observed until the end of Eocene, between northernmost Antarctic Peninsula, and southern South America-Australasia. This process, previously proposed by Askin (1989) has now a strong support and evidences.

5.2 Late Cretaceous Terrestrial Vertebrates of the James Ross Basin

Evidence of Late Cretaceous (Coniacian to Maastrichtian) Antarctic terrestrial faunas are documented exclusively in the NE flank of the Antarctic Peninsula, West Antarctica (Reguero and Gasparini 2007; Reguero et al. submitted) (Fig. 5.8). Late Cretaceous deposits of the James Ross Basin have been extensively surveyed by vertebrate paleontologists; most collections consist of marine reptiles (del Valle et al. 1977; Gasparini et al. 1984; Chatterjee and Small 1989; Martin et al. 2002; Novas et al. 2002; Martin 2006; Martin and Crame 2006; de la Fuente et al. 2010) and other marine vertebrates (Woodward 1908; Cione and Medina 1987; Grande and Chatterjee 1987; Richter and Ward 1990; Stahl and Chatterjee 1999, 2002; Kriwet et al. 2003; 2006). In contrast there have been fewer findings of Cretaceous terrestrial vertebrates, which correspond only to avian and non avian dinosaurs. The first Late Cretaceous terrestrial fossil vertebrates reported from Antarctica came from the James Ross Basin (Gasparini et al. 1987).

The Late Cretaceous non-avian and avian dinosaurs of the James Ross Basin, Antarctic Peninsula are numerically scarce and most of them are poorly known. There are specimens from Santonian (Molnar et al. 1996), Campanian (Olivero et al. 1991; Coria et al. 2008; Cerda et al. 2011), and Maastrichtian (Hooker et al. 1991; Rich et al. 1999; Case et al. 2000; Cambiaso et al. 2002) ages. The principal

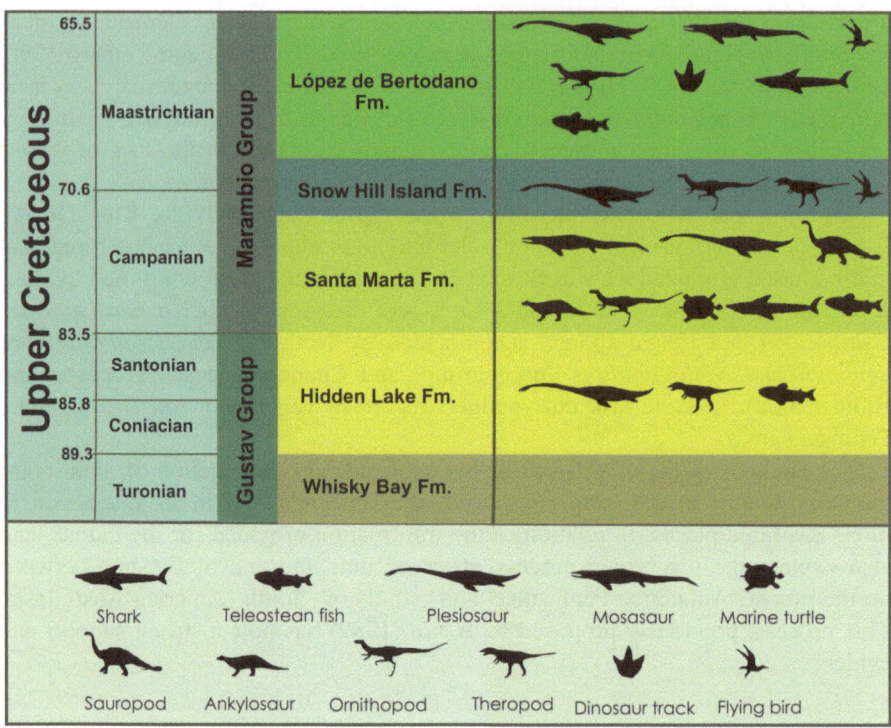

Fig. 5.8 Time scale, stratigraphy and vertebrate fossil record for the Late Cretaceous rocks in the James Ross Basin, Antarctic Peninsula, West Antarctica. Temporal and sedimentary units not to scale

reason for the paucity of the record of fossil dinosaurs and birds in the James Ross Basin is that the hosting sediments were deposited in shallow marine or coastal depositional settings. Therefore, any recovered specimens were originally derived from terrestrial environments and floated out from shore, to be preserved against high odds in a marine setting.

At least nine taxa of non-avian dinosaurs are now known: a megalosaur-like theropod (Molnar et al. 1996), a nodosaurid ankylosaur (Salgado and Gasparini 2006), a dromaeosaurid theropod (Case et al. 2007), hadrosaurs (Rich et al. 1999; Case et al. 2003), iguanodontids (Novas et al. 2002; Coria et al. 2008), a hypsilophodontid (Hooker et al. 1991), a large-bodied lithostrotian titanosaur (Cerda et al. 2012), and at least four avian dinosaurs have been reported or described from the Maastrichtian deposits of this basin. Additional non-avian dinosaur evidence from the same area is based upon the occurrence of Maastrichtian sauropod? footprints of Snow Hill Island (Olivero et al. 2007). The non-avian dinosaurs from West Antarctica seem to be remnants of a cosmopolitan dinosaur fauna more typical of the Campanian and Maastrichtian of Patagonia and other Gondwanan areas. Antarctic Late Cretaceous avian dinosaurs are rare and they are restricted to one gaviid (Chatterjee 1989), a charadriform (Case and Tambussi 1999), and

Fig. 5.9 Avian dinosaur from the Late Cretaceous of Vega Island, James Ross Basin, Antarctic Peninsula. *Vegavis iaai* Clarke, Tambussi, Noriega, Erickson and Ketcham 2005, MLP 93-I-3-1, holotype: **a** the half of the concretion preserving most of the skeleton of the holotype; **b** the other half of the concretion with few bones; **c** right humerus and **d** left tibiotarsus. Scale bar = 20 mm. Abbreviations: *c* coracoids; *cv* cervical vertebra; *f* femora; *h* humerus; *il* ilium; *p* pubis; *r* radius; *tmt* tarsometatarsus; *tv* thoracic vertebrae; *s* sacrum; *sc* scapula; *u* ulna

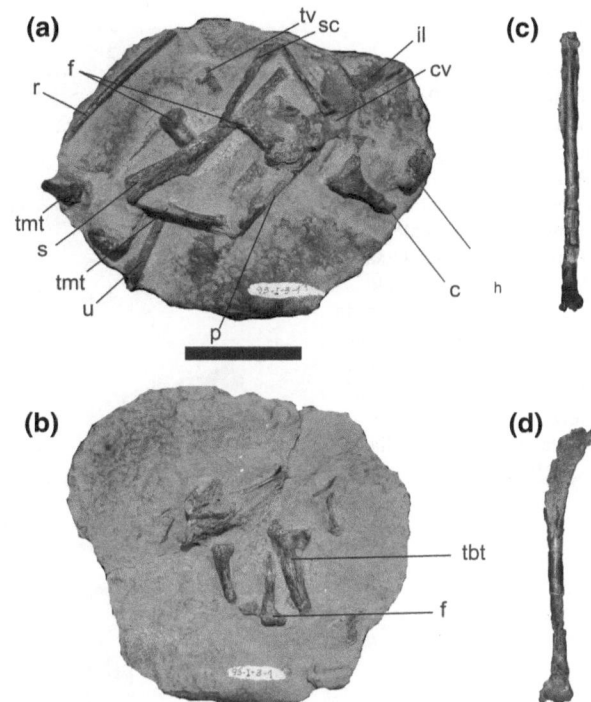

a basal Anseriformes, *Vegavis iaai*, which provided the first strong evidence for a basal part of the extant avian radiation in the Cretaceous (Clarke et al. 2005) (Fig. 5.9).

5.2.1 Non Avian Dinosaurs

The first dinosaur found in Antarctica, *Antarctopelta oliveroi* Salgado and Gasparini 2006, from the Gamma Member of the Santa Marta Formation on the northeastern James Ross Island (Gasparini et al. 1987), corresponds to a small-sized ankylosaur. The specimen MLP 86-X-28-1 is represented by a partial skeleton consisting of a fragment of left dentary with a tooth, cervical, dorsal and sacral vertebrae, metacarpals and several scutes (Fig. 5.10).

The James Ross Island ankylosaur was collected from shallow marine deposits (Fig. 5.11), associated with various marine invertebrates such as bivalves, gastropods and ammonites that indicate a late Early to basal Late Campanian age (Olivero et al. 1991; Olivero 2012).

Hooker et al. (1991) reported a partial skeleton of a large-bodied (4–5 m) ornithopod dinosaur from the Early Maastrichtian Cape Lamb Member of the Snow Hill Island Formation on Vega Island (Fig. 1.1). Numerous skeletal characteristics

Fig. 5.10 Non avian dinosaurs from Late Cretaceous of James Ross Island, Antarctic Peninsula. *Antarctopelta oliveroi* Salgado and Gasparini 2006. MLP 86-X-28-1, holotype, **a** posterior cervical vertebra in anterior view. Scale bar: 50 mm; **b** right metatarsal IV? in lateral view. Scale bar: 50 mm; **c** left dentary in occlusal view; **d** left dentary in medial view. Scale bar: 50 mm; **e** tooth II in lingual view. Scale bar: 20 mm; **f** distal end of metapodial. Scale bar: 50 mm

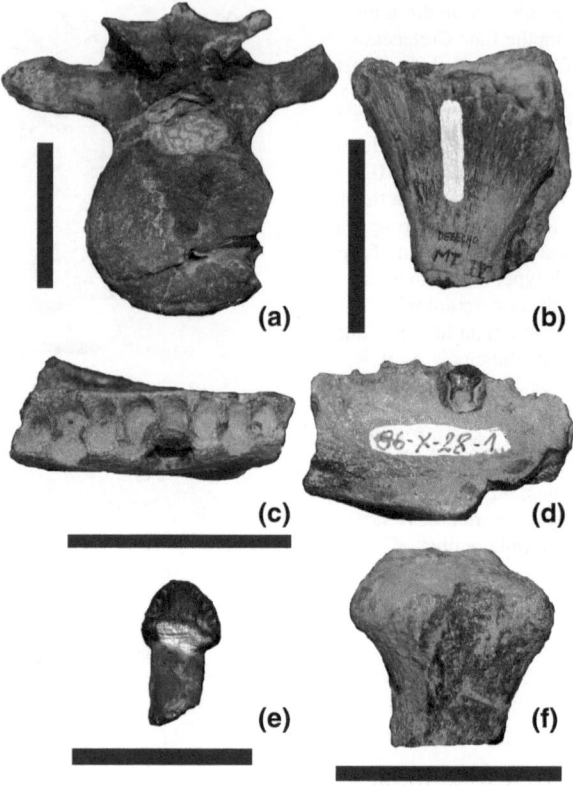

indicate that this Antarctic specimen is a member of Hypsilophodontidae, although some postcranial features also resemble those of dryosaurids (Iguanodontia) (Milner et al. 1992).

Also from the James Ross Island, but further to the north from the ankylosaur locality, the distal portion of a tibia of a theropod was recovered from older deposits, i.e., the Coniacian Hidden Lake Formation at Cape Lachman, James Ross Island (Molnar et al. 1996).

Rich et al. (1999) described a distal end of an ornistichian metatarsal (Fig. 5.12a–c), referred by the authors to a probable hadrosaur. It comes from the Late Maastrichtian unit Klb 9 of Macellari (1988) of the López de Bertodano Fm. on Seymour Island (Fig. 5.13).

Case et al. (2000) reported the presence of dental and pedal material of a hadrosaurine hadrosaur in the Late Maastrichtian Sandwich Bluff Member of the López de Bertodano Formation on Vega Island. The "Reptile Horizon", named for the numerous mosasaur and plesiosaur specimens recovered from this stratigraphic level, is in the upper third of the Sandwich Bluff Member (Crame et al. 1991; Pirrie et al. 1991; Martin 2006) (Fig. 5.14), Unit C (Olivero 1992) or K3 (Marenssi et al. 1992) of the López de Bertodano Formation. This member is

Fig. 5.11 Panoramic view of Santa Marta Cove, James Ross Island, Antarctic Peninsula. Lithostratigraphy of the Santa Marta Formation (Herbert Sound/Gamma members)

Fig. 5.12 Non avian dinosaurs from Late Cretaceous of James Ross Island, Antarctic Peninsula. Hadrosauriidae? MLP 96-I-6-2, distal end of metatarsal, **a** lateral view, **b** dorsal view, **c** ventral view. Scale bar: 50 mm; **d** Lithostrotian gen. et sp. indet. MLP 11-II-20-1, caudal vertebra centrum, right lateral view. Scale bar: 50 mm

Fig. 5.13 General view of the uppermost horizons of the López de Bertodano Formation at Seymour Island showing the K/T boundary between units Klb 9 (Late Maastrichtian) and Klb 10 (Danian) within the López de Bertodano Formation

composed of nearshore marine fine-grained, ferruginous, loosely consolidated sandstones that are late Maastrichtian in age (approximately 66–68 Ma), based on correlations of ammonite and palynological taxa (Crame et al. 1991; Pirrie et al. 1991; Olivero and Medina 2000; Olivero 2012).

During the 2006–2007 field season Judd Case and his field crew (including M.A.R.) collected vertebrate specimens from the lower portion of the Cape Lamb Member of the Snow Hill Island Formation (Crame et al. 2004) on The Naze, northern James Ross Island, (Fig. 1.1) consisting of a partial skeleton of a small-bodied theropod dinosaur (Case et al. 2007). The specimen includes a metatarsal II with a lateral expansion caudal to metatarsal III, a third metatarsal that is proximally narrow and distally wide, a metatarsal III with a distal end that is incipiently ginglymoidal and a second pedal digit with sickle-like ungula. These are all features diagnostic of a theropod that belongs to a family of predatory dinosaurs, the Dromaeosauridae (Fig. 5.15a). The Antarctic dromaeosaur exhibits additional diagnostic characters of Dromaeosauridae, i.e., Mt II with a lateral expansion caudal to Mt III; Mt III proximally narrow distally wide (Novas and Pol 2005). It is of great interest that new dromaeosaur species are being recovered from the mid-Cretaceous of South America; this fact, together with the retention of primitive characters in the Antarctic dromaeosaur, suggests a new biogeographic hypothesis for dromaeosaur distribution. Gondwanan dromaeosaurs are not North American immigrants into South America and Antarctica; rather they are the relicts of a cosmopolitan dromaeosaur distribution, which has been separated by the vicariant break up of Pangea and created an endemic clade of dromaeosaurs in Gondwana.

Fig. 5.14 Panoramic view of the Sandwich Bluff, Cape Lamb, Vega Island, Antarctic Peninsula showing stratigraphy of the Snow Hill Island (Early Maastrichtian Cape Lamb Member) and the López de Bertodano (Late Maastrichtian Sandwich Bluff Member) formations and the locations of the fossil localities bearing dinosaurs. VEG9303 is on a ridge, which divides two amphitheaters at the base of the Sandwich Bluff. Miocene volcanic beds belonging to the James Ross Island Volcanic Group caps Leal Bluff in the background

From similar-aged deposits on James Ross Island, portions of a hindleg and foot of an iguanodontid (MACN-19777) were recovered (Cambiaso et al. 2002; Novas et al. 2002). The exact stratigraphic and geographic location of this iguanodontid is uncertain, although field evidence suggests that it came from Fortress Hill next to Terrapin Hill in The Naze (Fig. 1.1), and thus belongs to the same lithostratigraphic unit as the dromaeosaur mentioned above.

Ornithopod remains have been also reported by Coria et al. (2007, 2008) from the environs of the *Antarctopelta* site (Fig. 5.11). Those include an isolated ungual phalanx (Coria et al. 2007) and an incomplete but semiarticulated and very informative skeleton of a small basal ornithopod (Coria et al. 2008) (Fig. 5.15b). This new form is currently under study and their phylogenetic relationships will be useful to propose some insights about the paleobiogeography of this fauna.

Recently, a vertebra of a sauropodomorph was recovered in shallow marine shelf deposits referred to the Late Campanian Santa Marta Formation in the northern part of James Ross Island (Cerda et al. 2011) (Fig. 5.12d). The size and morphology of the specimen allows its identification as a caudal vertebra (MLP 11-II-20-1) of a lithostrotian titanosaur. This discovery increases the fossil dinosaur richness of Antarctica and enhances our understanding of the global distribution of sauropod dinosaurs (Cerda et al. 2011). MLP 11-II-20-1 is the second sauropodomorph dinosaur recorded from Antarctica. The first one is a basal

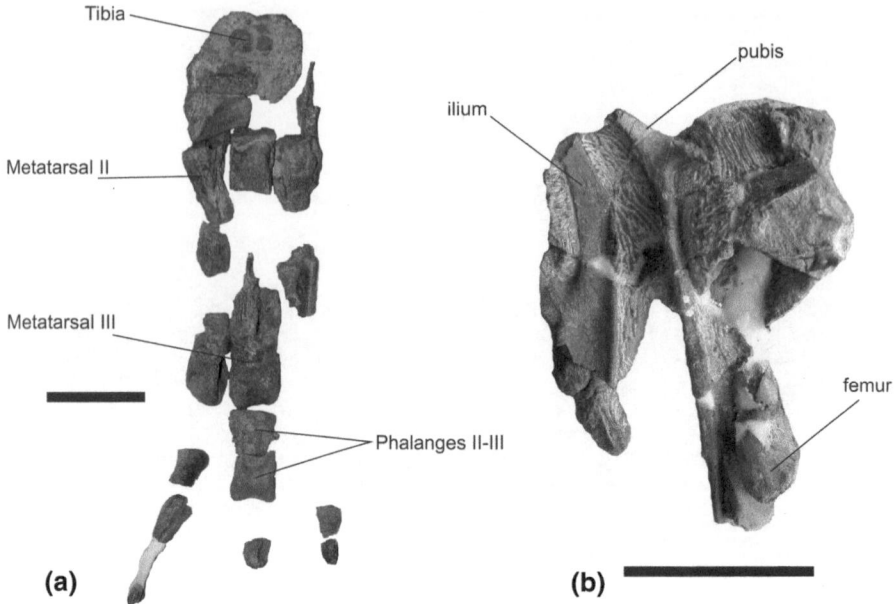

Fig. 5.15 Non avian dinosaurs from Late Cretaceous of James Ross Island, Antarctic Peninsula. **a** Dromaeosauridae gen et sp. indet., partial left hindlimb, ankle and foot; **b** Ornithopod gen. et sp.nov., MLP 08-III-1-1, associated pubis, ilium and femur of the skeleton. Scale bar: 100 mm

sauropodomoph dinosaur, *Glacialisaurus hammeri*, collected from the Beardmore Glacier region of the Central Transantarctic Mountains (Early Jurassic, Hanson Formation) (Hammer and Hickerson 1994; Smith and Pol 2007).

Finally, additional non-avian dinosaur evidence from the James Ross Basin is based upon the occurrence of footprints in southwestern part of the Snow Hill Island (Spath Peninsula) where aligned depressions interpreted as poorly preserved dinosaur footprints were reported by Olivero et al. (2007). These ichnites suggest a more extensive distribution of dinosaurs within the basin and southward, and this circumstance could be reflecting opportunistic dispersal of members of the group at the end of the Cretaceous among southern Gondwanan areas via Antarctica, as was previously proposed (Sampson et al. 1998), probably during periods of more favorable climatic conditions in the region.

5.2.2 Avian Dinosaurs

The first record of a Mesozoic Antarctic bird was a loon, referable to the modern family Gaviidae. The discovery of *Polarornis gregorii* was first announced unofficially by Sankar Chatterjee in 1989 and later in his 1997 book *The rise of birds*. The material consists on incomplete remains of one individual with poor

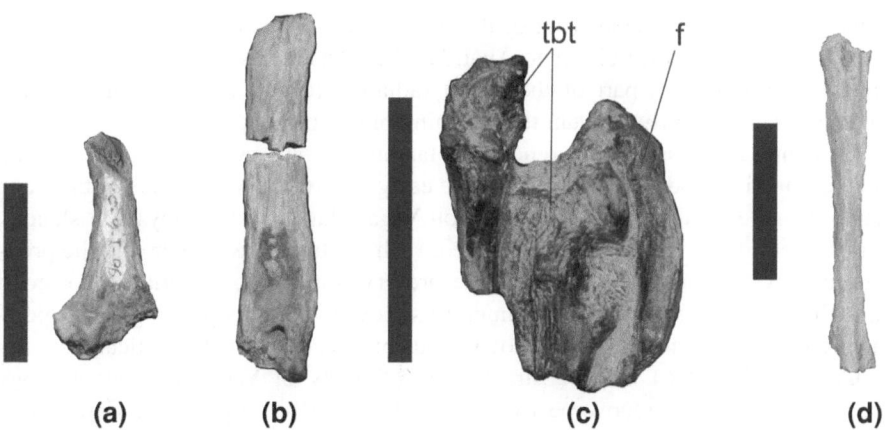

(a)   **(b)**   **(c)**   **(d)**

Fig. 5.16 Avian dinosaurs from the Late Cretaceous of Vega and Seymour islands, James Ross Basin, Antarctic Peninsula. **a** *Polarornis gregorii?* MLP 96-I-6-4, distal end of left femur. Scale bar: 50 mm; **b** *Polarornis gregorii?* MLP 96-I-6-3, distal end of left tibiotarsus. Scale bar: 50 mm; **c** *Polarornis gregorii* Chatterjee 1989, MLP 96-I-6-2, concretion containing the two tibiotarsi and the right femur of the same individual. Scale bar: 50 mm; **d** Charadriiform gen. et sp. indet. MLP 98-I-10-25, fragment of left tarsometatarsus. Scale bar: 20 mm. Abbreviations: *f* femur; *tbt* tibiotarsus

preservation (fragment of the bill and the adjacent distal part of the skull, parts of the otic region, four vertebrae, a small fragment of the sternum, a femur, and a proximal end of a tibiotarsus) found in rocks of the López de Bertodano Formation on Seymour Island. Chattherjee did not officially name the species until 2002. Additional material (MLP 96-I-6-2/3) possibly belonging to the same taxon is figured here for the first time (Fig. 5.16a). Chaterjee et al. (2006) reported a well-preserved skeleton of a new gracile species of a fossil loon *Polarornis* from the Early Cretaceous Sandwich Bluff Member of the López de Bertodano Formation on Vega Island. Although *Polarornis* is often claimed to be an ancestor of modern loons (Gaviiformes), its relationships are unclear (Tambussi and Acosta Hospitaleche 2007). In his work of 2004, Mayr (2004) pointed out that *Polarornis* differs from modern loons in some important features. *Neogaeornis wetzeli,* from the Upper Cretaceous of Chile, based on a scrap of tarsometatarsus, has been interpreted as the earliest fossil gaviiform (Lambrecht 1929, Olson 1992) that resembles the highly derived bone of modern loons (Mayr 2004).

A second Cretaceous bird, firstly referred to Presbyornithidae by Noriega and Tambussi (1995), was collected from Late Maastrichtian strata of Cape Lamb (Locality VEG 9303, Fig. 5.9 and 5.14), Vega Island. Significant additional preparation, x-ray computed tomography and histological analysis revealed the presence of an array of elements not previously known and important neornithinae features. It was described as a new species of Anseriformes, *Vegavis iaii,* by Clarke et al. (2005; 2006). The specimen (MLP 93-I-3-1) consists on a disarticulated partial postcranium preserved in the two halves of a concretion (Fig. 5.9a, b), right humerus and left tibiotarsus (out of the blocks) (Fig. 5.9c, d). *Vegavis* is assessed to be a basal anseriform closer to Anatidae than Anseranas and it is important as the best support for a part of the extant radiation (crown clade) of birds in the Cretaceous and directly relevant to the timing of this radiation.

Case and Tambussi (1999) reported a tarsometatarsus, MLP 98-I-10-25, of an undetermined charadriiform bird from the early Maastrichtian deposits of the Cape Lamb Member, Snow Hill Island Fm. on Vega Island. The diaphysis is slender, being much longer than wide. It is eroded, with the trochleas broken and the proximal epiphysis absent (Fig. 5.16d). The preservation of the material is not good enough to permit its positive assignment; its reexamination is pending. This specimen represents the most ancient neognathous bird recorded in Antarctica.

Case et al. (2006) in an abstract of the Society of Vertebrate Paleontology reported a left femur recovered from the Sandwich Bluff area of the Cape Lamb on Vega Island (Fig. 1.1) and its stratigraphic position places it near the base of the Sandwich Bluff Member and at an equivalent stratigraphic level to that of *Vegavis* described by Clarke et al. (2005). It shows striking similarities to modern cursorial predatory birds of South America (Seriemas, Cariamidae) and of Africa (Secretarybirds, Sagittariidae).

5.3 Paleogene Terrestrial Vertebrates of the James Ross Basin

Our knowledge of the Eocene land mammals of Antarctica has gradually increased since the time when Woodburne and Zinsmeister described the first specimens in 1984 (Fig. 5.17).

In spite of this, fewer than 20 land mammal species are known (Table 5.1), and our understanding of these communities does not come close to our knowledge of the Eocene faunas of South America. The composition of the La Meseta terrestrial fauna differs in several aspects from that of Eocene faunas from the rest of South America (Reguero et al. 2002; but see below), suggesting that the Antarctic Peninsula was faunally not a part of Patagonia by the early to middle Eocene. Seymour Island contains the only Eocene land vertebrate fauna known from Antarctica, and represents the southernmost part of the distribution of some Paleogene South American (mostly Patagonian) land mammal lineages. Paleogeographic reconstructions of the Antarctic Peninsula during the Eocene, utilizing paleomagnetic data collected on the continent itself, indicate a paleolatitude perhaps as far south as 63° (Lawver et al. 1992).

In addition to other terrestrial vertebrates, the richly fossiliferous Eocene sediments of Seymour island have yielded the first land mammal (Woodburne and Zinsmeister 1984), the first placental land mammal (Carlini et al. 1990), and the first record of extinct "South American native ungulates" or SANU (Bond et al. 1990) from Antarctica (Reguero et al. 2002; Bond et al. 2011). It has also yielded dung beetle brood balls (Laza and Reguero 1993), leaves, tree trunks, a flower, and

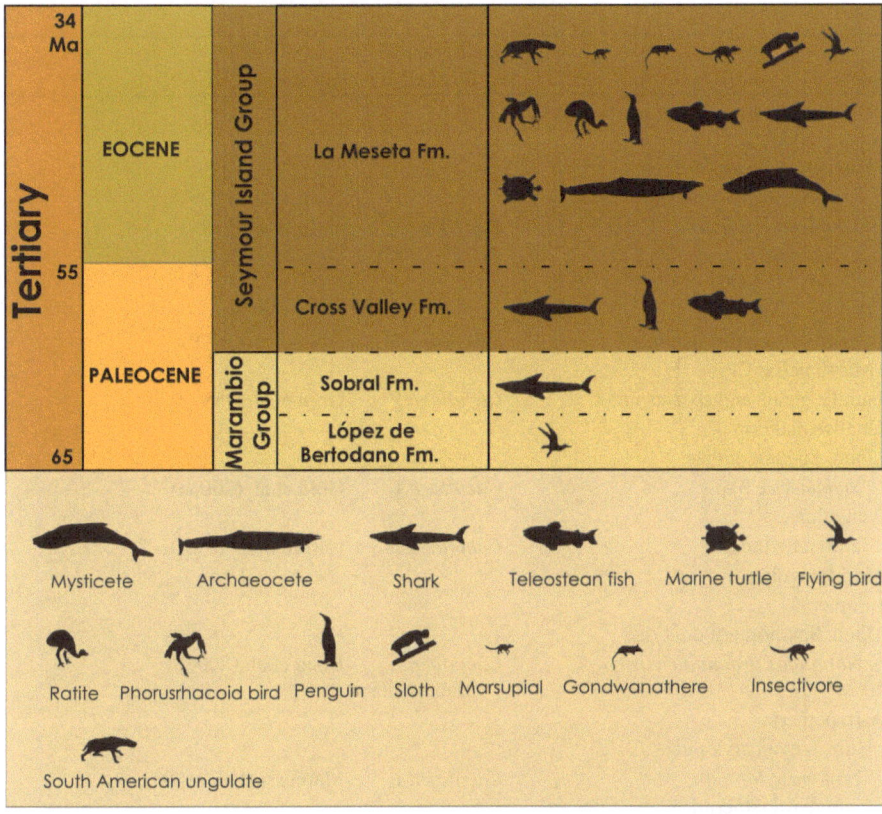

Fig. 5.17 Time scale, stratigraphy and vertebrate fossil record for the Paleogene rocks in the James Ross Basin, Antarctic Peninsula, West Antarctica. Temporal and sedimentary units not to scale

Table 5.1 Taxonomic list, stratigraphy, and references for the terrestrial and marine mammals from the Eocene La Meseta Formation of Seymour Island, Antarctic Peninsula

Taxon	Stratigraphy (Allomember)	Source
Mammalia		
Polydolopimorphia		
Fam. Prepidolopidae		
Perrodelphys coquinense	*Cucullaea* I	Goin et al. (1999)
Fam. Polydolopidae		
Antarctodolops dailyi	*Cucullaea* I	Woodburne and Zinsmeister (1984)
Antarctodolops mesetaense	*Cucullaea* I	Chornogubsky et al. (2009)
Microbiotheria		
Fam.? Microbiotheriidae		
Marambiotherium glacialis	*Cucullaea* I	Goin et al. (1995)
Fam. Woodburnodontidae		
Woodburnodon casei	*Cucullaea* I	Goin et al. (2007a, b)

(Continued)

Table 5.1 (Continued)

Taxon	Stratigraphy (Allomember)	Source
"Didelphimorhia" (basal Metatheria or Marsupialia)[a]		
Fam. Derorhynchidae		
Derorhynchus minutus	*Cucullaea* I	Goin et al. (1999)
Pauladelphys juanjoi	*Cucullaea* I	Goin et al. (1999)
Derorhynchidae, genus and species indet.	*Cucullaea* I	Goin et al. (1999)
Family indet.		
Xenostylos peninsularis	*Cucullaea* I	Goin et al. (1999)
?Marsupialia		
Family, genus and species indet.	*Cucullaea* I	Goin et al. (1999)
Gondwanatheria		
Fam. Sudamericidae		
Sudamerica? sp.	*Cucullaea* I	Goin et al. (2006a)
Xenarthra		
Tardigrada indet.	*Cucullaea* I	Carlini et al. (1990)
Meridiungulata		
Litopterna		
Fam. Sparnotheriodontidae		
Notiolofos arquinotiensis	*Cucullaea* I, Submeseta	Bond et al. (2006)
Astrapotheria		
Fam. Trigonostylopidae		
Trigonostylops sp.	*Cucullaea* I	Marenssi et al. (1994)
Fam. Astrapotheriidae		
Antarctodon sobrali	*Cucullaea* I	Bond et al. (2011)
"Insectivora"		
gen. et sp. indet.	*Cucullaea* I	MacPhee et al. (2008)
Mammalia *incertae sedis*		
gen. et sp. indet.	Acantilados	Vizcaíno et al. (1997a)
Cetacea		
Fam. Basilosauridae		
Zeuglodon sp.	Submeseta	Wiman (1905ab)
Zygorhiza sp.	Submeseta	Cozzuol (1988)
gen. et sp. nov	*Cucullaea* I	Buono et al. (2011)
Mysticeti		
Crenaticeti		
Llanocetus denticrenatus	Submeseta	Mitchell (1989)

[a]"Didelphimorphia" in its traditional concept and contents is not regarded as a natural group

diverse marine vertebrate and invertebrate faunas (Stilwell and Zinsmeister 1992; Torres et al. 1994; Gandolfo et al. 1998a, b).

The oldest dates in the formation, 52–54 Ma (based on $^{87}Sr/^{86}Sr$ isotopic ratios after Reguero et al. 2002), come from the 150 m level on the boundary between the Valle de las Focas and Acantilados allomembers of Marenssi and Santillana (1994). Consequently, the base of the La Meseta should be close to the beginning of the Eocene at 55.8 Ma (Gradstein et al. 2004). Most of the Early Eocene and later aged terrestrial mammalian groups found in West Antarctica have inferred ancestral lineages close to those found in mainland Patagonia (Reguero et al. 2002, Reguero and Marenssi 2010).

Among the Eocene terrestrial mammals from the La Meseta Formation, Seymour Island, the meridiungulate litoptern Sparnotheriodontidae and the marsupial family Polydolopidae were the dominant taxa. The former was not a dominant element in the much larger Paleogene associations elsewhere in South America (Reguero et al. 1998, 2002). In turn, the latter are usually the dominant element among the Patagonian metatherian taxa of Paleocene-Eocene age. A quite abundant "group" in the terrestrial WANT fauna is a suite of "opposum-like" metatherians (Goin et al. 1999).

5.3.1 Gondwanatheres

Gondwanatherians are a group of extinct, non-therian mammals of still uncertain affinities (Pascual et al. 1999). The most advanced members of this clade, the Sudamericidae, represents the first attempt among mammals towards the acquisition of high-crowned (hypsodont) cheek-teeth. Molariforms of sudamericids have a flat occlusal surface, which is divided into transverse lophs—a unique feature among mammals that lived on Gondwanan continents during the Cretaceous and Paleocene. A small, rodent-like dentary fragment (MLP 95-I-10-5) preserving part of the incisor was recovered from Ypresian beds of the *Cucullaea* I Allomember (IAA 1/90 locality) (Fig. 5.18); both the incisor enamel structure and the mandibular morphology suggest close affinities with the sudamericid *Sudamerica ameghinoi* from the Selandian of Patagonia (Goin et al. 2006a). For the sudamericids, Koenigswald et al. (1999) inferred a semi-aquatic and perhaps a burrowing way of life, similar to that of living beavers.

Outside of West Antarctica, Cretaceous and Paleogene gondwanatherian mammals have been reported from several Gondwanan terranes (Fig. 5.19. The presence of gondwanatherians in Antarctica comfirms the Southern Hemisphere distribution of the group; notwithstanding, the fragmentary nature of the single specimen recognized from the Antarctic Peninsula prevents further comments on the evolution of this group on this continent.

5.3.2 Metatherians

Compared to other mammalian lineages, metatherians comprise the most diverse group so far recorded from levels of La Meseta Fm in the Antarctic Peninsula. Eight species have been described from these levels, referable to the "Didelphimorphia" (three species), Microbiotheria (two species), and

Fig. 5.18 Antarctic
sudamericid genus et species
indet., cf. *Sudamerica
ameghinoi* Scillato Yané and
Pascual 1984, MLP 95-I-
10-5, an anterior portion of
a left dentary showing the
enlarged, rodent-like incisor
partially preserved: **a** dorsal,
b anterior and **c** lateral
views. References: *i*, incisor;
mf, mental foramen. Scale
bar = 1 mm

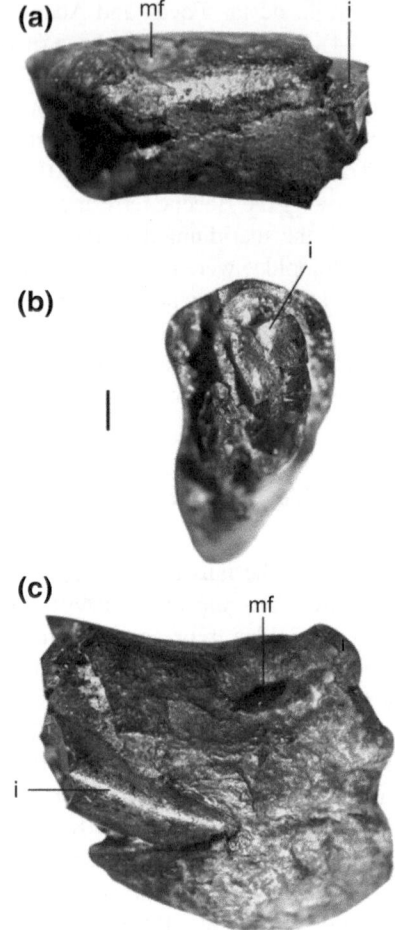

Polydolopimorphia (three species; see Woodburne and Zinsmeister 1984; Goin et al. 1999, 2007a; Chornogubsky et al. 2009). If an edentulous mandible not referable to any of the above, and a badly worn lower molar of an indeterminate Derorhynchidae are included (see Goin et al. 1999), the list of metatherian taxa can be considered as composed of at least ten species (see Table 5.1). This is, by far, the largest diversity among mammals in the Eocene of Antarctica. In specimen numbers, species of *Antarctodolops* (Polydolopidae) are, with 22 known fragments (Chornogubsky et al. 2009), the most abundant mammals from the La Meseta Fm. All other metatherian taxa known from this formation were recognized on the basis of less than a dozen specimens.

"Didelphimorphia" As traditionally conceived, i.e., including a large array of taxa ranging from "Alphadontidae" to Didelphidae, "Didelphimorphia" cannot be regarded as a natural group. Recently Horovitz et al. (2008) proposed a more restricted concept of Didelphimorphia, including only peradectids and didelphids. Thus, many lineages of basal metatherians can be excluded from it. Derorhynchidae is one of these groups.

Fig. 5.19 Southern Hemisphere continents showing fossil localities that include gondwanatherians, or presumed gondwanatherians. References (from older to younger): *1* Cretaceous (*s.l.*) of Tanzania: unnamed ?Gondwanatheria ("Red Sandstone Group"; Krause et al. 2003); *2* Late Cretaceous (Maastrichtian) of Madagascar: *Lavanify miolaka*, Sudamericidae (Maevarano Formation; Krause et al. 1997); *3–6.* Late Cretaceous (Maastrichtian) of India: *Bharattherium bonapartei*, Sudamericidae (Deccan intertrappean sediments; Prasad et al. 2007, Wilson et al. 2007 [*3* Gokak, *4* Naskal, *5* Kisalpuri, and *6* Bacharam localities]; *7–9.* Late Cretaceous (Campanian–Maastrichtian; Alamitian Age) of northern Patagonia, Argentina: *Ferugliotherium windhauseni* (Ferugliotheriidae), *Trapalcotherium matuastensis* (Ferugliotheriidae), and *Gondwanatherium patagonicum* (Sudamericidae) [*7* Los Alamitos (Los Alamitos Formation; Bonaparte 1986), *8* Cerro Tortuga (Allen Formation; Rougier et al. 2009), *9* La Colonia (La Colonia Formation.; Pascual and Ortiz-Jaureguizar 2007: Figs 4.11, 4.12)]; *10* medial Paleocene (early Selandian) of central Patagonia (Salamanca Formation, Punta Peligro locality; Scillato-Yané and Pascual 1985); *11* early-middle Eocene of northern Patagonia, Argentina: *Greniodon sylvaticus* (Family indet) (Volcanic-Pyroclastic Complex of the Middle Chubut River, La Barda locality; Goin et al. 2012); *12* early-middle Eocene of the Antarctic Peninsula: Sudamericidae, gen. et sp. indet., "cf. *Sudamerica ameghinoi*" (La Meseta Formation, locality IAA 1/90; Goin et al. 2006a: 136); *13* middle-late Eocene of Perú: "Sudamericidae", undescribed specimens and taxa (Contamana locality, Yahuarango Formation; Antoine et al. 2011 [electronic supplementary material http://dx.doi.org/10.1098/rspb.2011.1732]); *14* ?late Eocene-early Oligocene of Perú:?Gondwanatheria, gen. et sp. indet. (Santa Rosa locality, Yahuarango Formation; Campbell et al. 2004; Goin et al. 2004)

Derorhynchids are characterized by their relatively small size, elongated snouts, and a molar dentition with specializations towards insectivorous to faunivorous habits. Dentaries are long and low at their mesial half. The first and second lower incisors are much larger than the third and fourth ones; there are large diastemata between c-p1 and p1-p2; upper molars have a large stylar cusp B which sometimes is located somewhat posteriorly as compared to the generalized metatherian pattern, the centrocrista is deeply "V"-shaped, and the protocone is antero-posteriorly short; lower molars have short talonids, reduced paracristids and tall, spire-like entoconids. The largest of the Antarctic derorhynchids is

Pauladelphys juanjoi (Fig. 5.20e–h), a quite derived species characterized by a deeply invasive centrocrista, very high and laterally compressed entoconid, and a very large stylar cusp B. *Derorhynchus minutus* (Fig. 5.17c, d) is the smallest known species of the genus, species of which are also present in latest Paleocene-early Eocene (Itaboraian SALMA) levels of southeastern Brazil and central Patagonia (Oliveira and Goin 2011). It is distinct in having a somewhat reduced protoconid and a relatively wide talonid in its lower 8and only known) molars. A third taxon of derorhynchid is possibly represented by a songle, much worn lower molar, probably a m2 or m3 (Fig. 5.20a, b). It is slightly larger than *D. minutus* and has a very deep metacristid notch.

Xenostylos peninsularis [the genus was originally named *Xenostylus* by Goin et al. (1999); preoccupied by *Xenostylus* (Insecta, Coleoptera); see Goin (2007a)] was first included among the Antarctic derorhynchids. However, since its initial description it was noted the peculiar morphology of the upper molars of this taxon (Fig. 5.20i-j): "The most distinctive features of *Xenostyl[o]s peninsularis* are its low crown height, very wide trigon basin, deep V-shaped centrocrista, para- and metaconule almost merged at the base of para- and metacone, a unique notch that divides the postparacrista, and the anterior "shifting" of StD in relation to most ameridelphian marsupials (…) [It] differs strikingly from other Antarctic derorhynchids, microbiotheriids, protodidelphids [lapsus for prepidolopids], and polydolopines, all of which seem to have more direct affinities with known South American taxa" (Goin et al. 1999: 348). Because of these same reasons here we opt not to include them among the Derorhynchidae, but instead regard this unique morphology as pertaining to a still undetermined lineage (Family indet.) of Metatheria.

Microbiotheria Microbiotherians were first recognized in strata of the La Meseta Fm from an edentulous dentary (Goin and Carlini 1995). Posteriorly Goin et al. (1999) described another dentary bearing a last lower molar (*Marambiotherium glacialis*; Fig. 5.21d, e). More recently Goin et al. (2007a) recognized a new taxon, *Woodburnodon casei* (Woodburnodontidae; Fig. 5.21a, b), based on a relatively large upper molar. *Marambiotherium* shares with the South American *Mirandatherium* (Itaboraian SALMA) a few features, such as a poorly reduced m4, very low hypeconid, and metaconid which is set on line with the protoconid. Except for the latter, none of these features are present in more modern (mid- Eocene onwards) South American microbiotheriids; in consequence, it could be the case that both Antarctic taxa, as well as *Mirandatherium*, are in fact referable to the Woodburnodontidae. As no lower molars of *Woodburnodon* have yet been discovered, this last alternative remains untested.

Polydolopimorphia   *Perrodelphys   coquinense*   (Bonapartheriiformes, Prepidolopidae) is another case of a taxon whose unique morphology including both generalized and derived features prevents further comparisons with other members of the family. It was recognized on the basis of a single specimen, a lower molar (m?1; Fig. 5.21k–m), which, as it happens with South American prepidolopids, has a much reduced paraconid which is close to the protoconid. In the case of *Perrodelphys*, however, this condition is extreme: the paraconid appears as an almost unrecognizable cuspule placed on the mesio-lingual slope of the

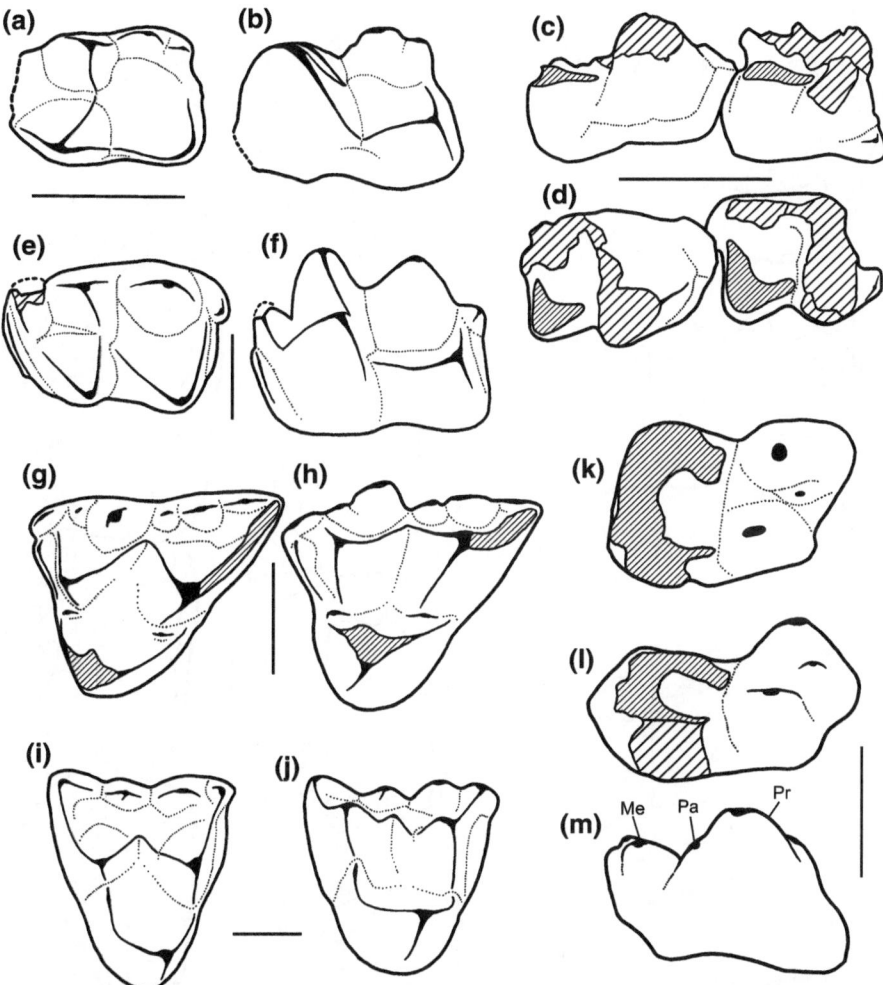

Fig. 5.20 Antarctic metatherians from the La Meseta Fm. (Antarctic Peninsula, WANT; early to middle Eocene). **a, b** Derorhynchidae, gen. et sp. indet.; specimen MLP 94-III-15-11, an isolated left lower molar in occlusal (**a**) and dorsolabial (**b**) views. **c, d** *Derorhynchus minutus* (Derorhynchidae); holotype, MLP 96-I-5-44; partially broken m2-m3 in obliquely labial (**c**), and occlusal (**d**) views. **e, f** *Pauladelphys juanjoi* (Derorhynchidae); holotype, MLP 95-I-10-2, an isolated right lower molar (m?3) in occlusal (**e**) and labial (**f**) views. **g, h** *Pauladelphys juanjoi*; referred specimen MLP 96-I-5-45, a left M?1 in occlusal (**g**) and lingual (**h**) views. **i, j** *Xeno-stylos peninsularis* (family indet); holotype, MLP 94-III-15-10, an isolated right upper molar in occlusal (**i**) and lingual (**j**) views. **k–m** *Perrodelphys coquinense* (Polydolopimorphia, Bonapar-theriiformes, Prepidolopidae); holotype, MLP 96-I-5-11; an isolated left lower molar (m?1) in occlusal (**k**), lingual (**l**), and anterior (**m**) views. References: Me, metaconid; Pa, paraconid; Pr, protoconid. Scale bar = 1 mm

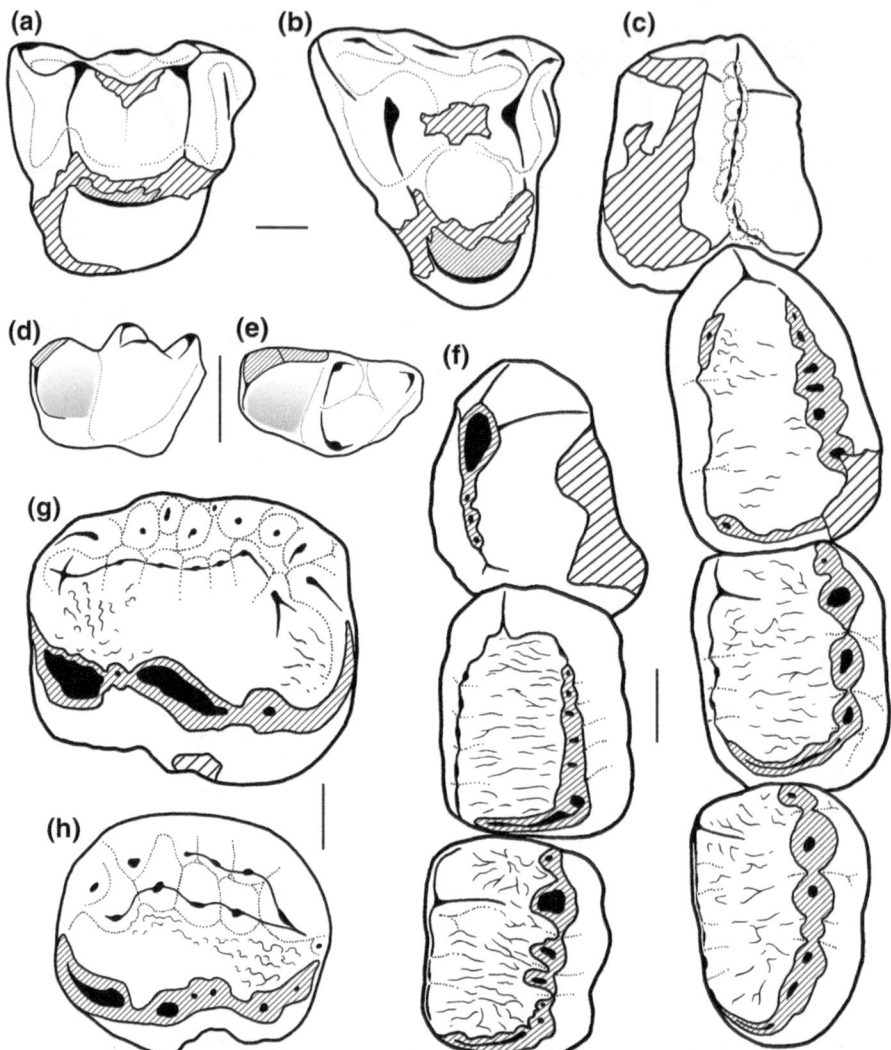

Fig. 5.21 Antarctic metatherians from the La Meseta Fm (Antarctic Peninsula, WANT; early to middle Eocene). **a, b** *Woodburnodon casei* (Microbiotheria, Woodburnodontidae); holotype, MLP 04-III-1-2, an isolated upper right molar (M2 or M3) in lingual (**a**) and occlusal (**b**) views. **d, e** *Marambiotherium glacialis* (Microbiotheria, fam. indet.): detail of the holotype, MLP 95-I-10-1, showing the last molar (m4) in dorsolabial (**d**) and occlusal (**e**) views. **c, f–h** species of *Antarctodolops* (Polydolopimorphia, Polydolopiformes, Polydolopidae). **c** *Antarctodolops mesetaense*; holotype, MLP 96-I-5-12, a right dentary with p3-m3 in occlusal view. **f** *Antarctodolops dailyi*; MLP 94-III-15-254, a right dentary with p3-m2 in occlusal view. **g** *Antarctodolops mesetaense*; MLP 95-I-10-4, an isolated left M1 in occlusal view. **h** *Antarctodolops dailyi*; MLP 88-I-1-4, an isolated right M1 in occlusal view. Scale bar = 1 mm

protocone. As it happens in *Prepidolops* from Northwestern Argentina, it could be the case that the Antarctic molar represents an m1. In *Prepidolops*, and due to the very large size of p3 (whose posterior edge overlaps the trigonid mesialmost edge), the paraconid of the m1 is posteriorly shifted and reduced. In *Perrodelphys* this feature is extreme, implying that the p3 was, proportionally, even larger. In turn, it is noteworthy that the talonid of *Perrodelphys coquinense* is quite generalized in aspect.

Antarctic polydolopids (Polydolopimorphia, Polydolopiformes) are referable to a single genus: *Antarctodolops* (Woodburne and Zinsmeister 1982, 1984). A recent review of all known specimens by Chornogubsky et al. (2009) led to the recognition of two species: *A. dailyi* (Fig. 5.21f, h) and *A. mesetaense* (Fig. 5.21c, g). An isolated specimen could possibly be referred to a third, still undescribed species of this same genus. As mentioned, *Antarctodolops* is the most abundant mammal from the Eocene of the Antarctic Peninsula. Both species of the genus are highly specialized towards omnivorous or omnivore-frugivorous feeding habits. Molars are relatively flat, with large masticatory surfaces, multicusped (the upper ones with several rows of cusps), the occlusal faces show wrinkled enamel, and the P2-3 as we ll as the p3 are large and plagiaulacoid-like. The clearly do not represent a generalized pattern for the rest of the family, as their molar morphology is clearly derived.

Affinities of West Antarctic metatherians The somewhat limited diversity of West Antarctic metatherians offers a mixed set of indicators when its taxon-by-taxon affinities are considered. First, there are taxa that, at least at the generic level, are shared with faunas of southern South America. This is clearly the case of *Pauladelphys*; *P. juanjoi* is closely related to a similarly sized, still unnamed, species from the early-middle Eocene (Ypresian-Lutetian boundary) of Paso del Sapo in western Patagonia (Tejedor et al. 2009). *P. juanjoi* is slightly more derived than the Patagonian species; for instance, it has a less reduced paraconid in the lower molars, while in the uppers the stylar cusp B is less posteriorly displaced. A second species representing this same case is *Derorhynchus minutus*; a closely related species is currently being described (F. Goin and A. Forasiepi, unpublished data) from the mid- Paleocene (early Selandian) of Punta Peligro, in eastern Patagonia. The age of the Peligran species argues in favor of a very early (Late Maastrichtian?, Danian?) arrival of South American metatherians into West Antarctica. A third example of this case may be that of the microbiotherian *Marambiotherium glacialis*, whose m4 shows derived features in common with the late Paleocene-early Eocene *Mirandatherium alipioi*, from southeastern Brazil.

In the second place, there are taxa that, though related to South American metatherians, are exclusive to West Antarctica. This is the case with the species of the polydolopid *Antarctodolops*, a genus exclusive to WANT. Species of *Antarctodolops* are not generalized but quite derived, and share some features in common with those of *Polydolops* and *Amphidolops* (Chornogubsky et al. 2009). Taking in account that the generalized polydolopimorphian *Cocatherium lefipanum*, from the early Danian of western Patagonia, could be basal to the Polydolopiformes radiation, it could be argued that the arrival of polydolopids in West Antarctica was a Paleocene, post-Danian event.

Finally, a third category of taxa is that of the higly derived, endemic species *Xenostylos peninsularis* (Family indet.) and *Perrodelphys coquinense* (Prepidolopidae). *Xenostylos* cannot be related to any particular lineage of South American, "opossum-like" metatherian. In turn, even though referrable to the Prepidolopidae, *Perrodelphys* is unique in its combination of generalized and derived features. Both species argue in favor of: (1) an early arrival to WANT of their ancestors (late Maastrichtian-Danian) from southern South America, and (2) some type of isolation mechanisms already active by early Paleogene times.

In summary, the diverse degree of specializations showed by metatherians from the La Meseta Fm favor the hypothesis of several waif dispersals from SAM to WANT instead of a single event. Also, that at least some degree of local distinction between southernmost SAM on one side, and of WANT on the other, within a general Weddellian terrestrial province, was already established at least since the mid- to late Paleocene.

Close phylogenetic affinities between South American and Australian metatherian taxa have already been argued (e.g., Sigé et al. 2004). Here we propose a few other hypotheses concerning Antarctic and Australasian metatherians, to be tested in future analyses:

1. *Xenostylos peninsularis* may prove to be basal to the Dasyuromorphian radiation. Some peculiarities of the upper molar morphology of this species, as the anteriorly shifted stylar cusp D, argue in favor of this relationship.
2. Derorhynchids may prove to be basal to the Peramelemorphian radiation.
3. Some lineages of Bonapartheriiform polydolopimorphian marsupials may prove to be basal to the Vombatiform (Diprotodontia) radiation. This implies that the Diprotodontia in their traditional concept is not a natural group.
4. Polydolopiform polydolopiformes are not ancestral to any lineage of Australasian marsupials.

5.3.3 Xenarthrans

During the 88/89 Antarctic field season, Argentinean palaeontologists recovered the first placental mammal from Antarctica (Carlini et al. 1990) at locality DPV 6/84. It was first identified as a megatherioid (Tardigrada) but new material collected from Colhuehuapian beds of Patagonia and referred to the Myrmecophagidae (Vermilingua) by Carlini et al. (1992) led these authors to reconsider the taxonomic and identification of the Antarctic specimen as ?Tardigrada or ?Vermilingua. This small-sized Antarctic sloth (ca. 10 kg) is considered to have been semi-arboreal and mainly folivorous (Vizcaíno et al. 1998). The trunk-climbing ability (scansoriality) of this form is difficult to assess, even though the strong, laterally compressed claw suggests this ability (Fig. 5.22a).

An isolated and incomplete tooth, discovered in sediments of Middle Eocene La Meseta Fm. on Seymour Island, was previously interpreted as pertaining to a

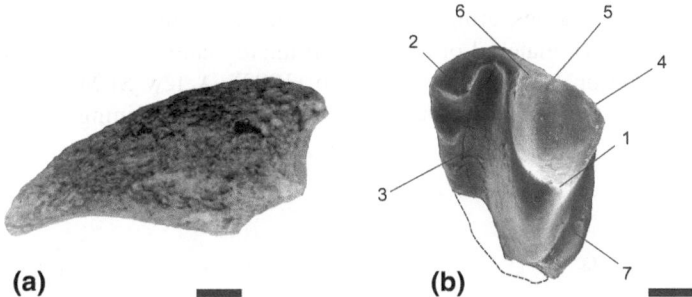

Fig. 5.22 Seymour Island xenarthran and "insectivore", *Cucullaea* I Allomember, La Meseta Formation: **a** ?Tardigrada or ?Vermilingua, MLP 88-I-1-95, gen. et sp. indet, ungueal phalanx species indet., Scale bar = 1 cm; **b** "Insectivora" gen. et sp. indet., MLP 91-II-4-3, right?, upper molar (M3?) in mesio occlusal view. References: *1* zalambdocone, *2* mesiostyle (parastyle), *3* precingulum, *4* distostyle (metastyle), *5* ectoflexus, *6* stylocone, and *7* postcingulum. Scale bar = 1 mm

sloth (Tardigrada). The specimen as preserved is composed of dentine, as in sloths and tooth-bearing xenarthrans generally. However, characters associated with the dentinal histology of definite sloths are either not represented on the Seymour tooth, or depart considerably from tardigradan and even general xenarthran models according to new observations presented by MacPhee and Reguero (2010). On the basis of histological criteria, the Seymour Island tooth cannot be shown positively to be tardigradan; it may not even be xenarthran. Further progress with establishing its relationships will depend on the recovery of more (and better) specimens. For the moment, it is best attributed to Mammalia *incertae sedis* (MacPhee and Reguero 2010).

5.3.4 *"Insectivora"*

Among the most bizarre terrestrial mammals of the Eocene of Seymour Island there is a minute "insectivorous"?placental, represented by an isolated tooth; its relationships are still undetermined. It was originally described as a "bat/insectivore" in affinity (Goin and Reguero 1993); however, the tooth is currently referred to Mammalia *incertae sedis*. The occlusal surface morphology and disposition of the roots of specimen MLP 91-II-4-3 (Fig. 5.22b) match well with a left maxillary molar, possibly the last one (MacPhee et al. 2008). The crown morphology of this tooth is characterized by a V-shaped crest, somewhat similar to that seen in the "zalambdomorph" molar occlusal pattern. Dentally, zalambdodont mammals generally have a large, central cusp in the upper molars from which two buccally extending ridges compose a relatively large stylar region and form a "V"-shape in occlusal view.

Unfortunately, and obscuring further more any possible comparison with other taxa, the only known material of this indeterminate mammal was lost after its preliminary description by Goin and Reguero (1993). A few SEM micrographs of specimen MLP 91-II-4-3 in various views are the sole remaining evidences of its existence.

5.3.5 Litopterns

The terrestrial vertebrate fauna from the *Cucullaea* I Allomember is unusual in being dominated by the large-sized and bizarre sparnotheriodontid litoptern *Notiolofos arquinotiensis* [the genus was originally named *Notolophus* by Bond et al. (2006); preoccupied by *Notolophus* Germar 1812 (Lepidoptera: Lymantriidae); see Bond et al. (2009)] (Fig. 5.23). This is not the case in the Eocene Patagonian fossil record where the terrestrial vertebrate faunas known from Vacan (middle Eocene), Barrancan (late Eocene) and Mustersan (late Eocene) are dominated by SANU orders and other ungulate groups such as kollpanines and Didolodontidae "archaic ungulates" and non-sparnotheriodontid litopterns.

Notiolofos could browse, stripping off twigs and saplings from evergreen trees even during winter months (Vizcaíno et al. 1998). Sparnotheriodontids share a number of dental characteristics that may be interpreted as adaptations to forested habitats (Reguero et al. 1998). As Marenssi et al. (1994) pointed out, the striking features of these mammals are brachyodonty and the particular structure of the enamel (vertically oriented Hunter-Schreger bands). Brachyodonty is associated with browsing herbivores that are adapted to forest habitats (Janis 1984). No postcranial information is available for the Antarctic meridiungulates, but information from the nearest relatives (all of them fossil) can be used to suggest locomotor adaptation. Both astrapotheres and sparnotheriodontids were medium to large ground mammals with restrictions in their mobility of limb articulations (presence of wrist and ankle joints that restrict lateral movement), as well as presence of hooves and reduction of digits in related taxa. In addition, the body size of the Antarctic sparnotheriodontid (395–400 kg) indicates that it was the largest herbivore living in Antarctica at this time (Vizcaíno et al. 1998). Evidently the large size of this herbivore favored the exploitation of leaves because longer residence time in the gut for bacterial fermentation is required to obtain sufficient nutrients from leaves. Also, it is accepted that large herbivores tend to feed more or less continuously on a wide range of plant parts. Low metabolic rate permits large herbivores to derive energy from cellulose by retaining it in the gut for long periods of microbial fermentation (Janis 1976). Based on dental morphology, astrapotheres and sparnotheriodontids probably were hindgut fermenters like non-ruminant artiodactyls and perissodactyls (Fortelius 1985). Astrapotheres and litoptern sparnotheriodontids also have teeth with vertical Hunter-Schreger bands. Fortelius (1985) indicates that a number of lophodont ungulates have convergently evolved vertically oriented Hunter-Schreger bands, a modification that involves the mode

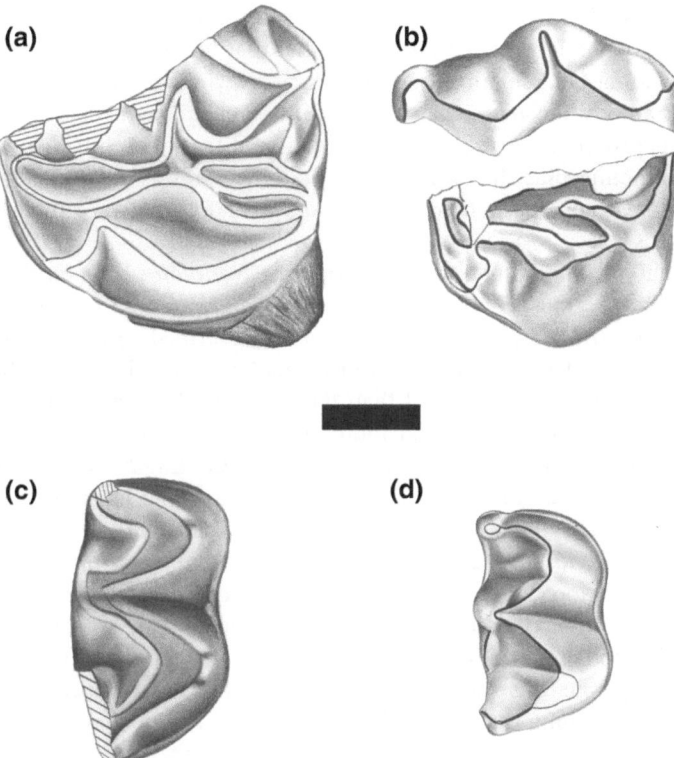

Fig. 5.23 *Notiolofos arquinotiensis* Bond et al. 2006, occlusal views of upper molars: **a** MLP 95-I-10-6, left M3, holotype, Submeseta Allomember of La Meseta Formation; **b** MLP 90-I-20-1, left M1 or M2, *Cucullaea* I Allomember of La Meseta Formation. Occlusal views of lower molars: **c** MLP 01-I-1-1, right m3, *Cucullaea* I Allomember of La Meseta Formation; **d** MLP 91-II-4-1, occlusal view of right p4, *Cucullaea* I Allomember of La Meseta Formation. Scale bar = 5 mm

of prism decussation and three-dimensional arrangement of the bands. This has been interpreted as an adaptation to resist cracking when the enamel edges are loaded in a direction away from the supporting dentine (Boyde and Fortelius 1986).

Although notoungulates were the most diverse (morphologically as well as taxonomically) and successful of the meridiungulates they are notably absent in West Antarctica. One of the most important radiations of notoungulates in Patagonia occurred during the late Paleocene-early Eocene. As we suppose that no barrier to dispersal existed between Patagonia and Antarctic Peninsula during the Paleocene, the absence of this group in Antarctica could be explained by suggesting that the La Meseta fauna is composed only of those taxa that were able to adapt to cooler conditions. If the notoungulates migrated southward into the Antarctic Peninsula during the late Paleocene, they presumably became extinct prior to the deposition of the La Meseta Formation (late early Eocene-late Eocene). Although it is certainly possible that notoungulates were present at La Meseta fauna but remain unsampled, the lack of specimens referable to this clade from among the more than 80 identified specimens at least speaks to the rarity of this group, if they existed in this region at that time; it is difficult to envision a taphonomic bias that would preserve other closely-related and similarly-sized ungulates (i.e., sparnotheriodontids and astrapotheriids) to the exclusion of notoungulates.

The Antarctic sparnotheriodontid litoptern, *Notiolofos arquinotiensis* (Bond et al. 2006), previously referred to *Victorlemoinea* (= *Sparnotheriodon*) by Bond et al. (1990), has close affinity with an undescribed species from Patagonia (Goin et al. 2000). The new taxon from Laguna Fría, Paso del Sapo, Chubut province, Patagonia, is more advanced than the primitive *Victorlemoinea prototypica* from the Itaboraian of Brazil, and more similar in size to *V. labyrinthica* from Cañadón Vaca (Vacan Age), Patagonia. However, the Laguna Fría taxon exhibits intermediate character states, i.e., more reduced hypocone in the first molars and shorter lingual crest of the metaconid, that show a close affinity with *Notiolofos arquinotiensis* (Bond et al. 2006). The record of *Notiolofos arquinotiensis* in the La Meseta Formation can be traced from the Campamento Allomember (Ypresian, Early Eocene) to the Submeseta Allomember (Priabonian, Late Eocene). In the *Cucullaea* I Allomember (Ypresian, Early Eocene), when the climatic conditions were cool-temperate, this species was common. Its last record occurs in the highest horizon of the Submeseta Allomember, dated by Dingle and Lavelle (1998) at ~34.2 Ma.

5.3.6 Astrapotheres

The occurrence of astrapotheres in La Meseta Formation was first reported by Bond et al. (1990). Hooker (1992) described a tooth fragment from a locality in northern Seymour Island as probably belonging to an astrapothere. The specimen BMNH BAS M2584 is in size, crown height, concave labial wall of the ectoloph, and postmetaconule crista similarly directed toward the metastyle, not converging with the postcingulum more similar to *Notiolophos arquinotiensis* than to an astrapothere (Bond et al. 2011). Marenssi et al. (1994) described and figured additional fragmentary material from the same horizon. Bond et al. (2008) pointed out that the Antarctic astrapothere is reminiscent of the Patagonian genus *Trigonostylops* but represents a different new taxon. Recently, Bond et al. (2011) described a tooth of a new genus and species, *Antarctodon sobrali* (Fig. 5.24), recovered from the *Cucullaea* shell bank (Telm 4 of Sadler 1988). This tooth shows sufficient distinctiveness from species of *Trigonostylops* and all other known astrapotheres to be interpreted as a new taxon. At present these autbors regard that the record of order Astrapotheria in West Antarctica should be considered as limited to astrapotheres not proximally related to *Trigonostylops*.

Fig. 5.24 *Antarctodon sobrali* Bond et al. 2011, MLP 08-XI-30-1, holotype, right p4 or m1 in **a** occlusal and **b** labial views. Scale bar = 5 mm

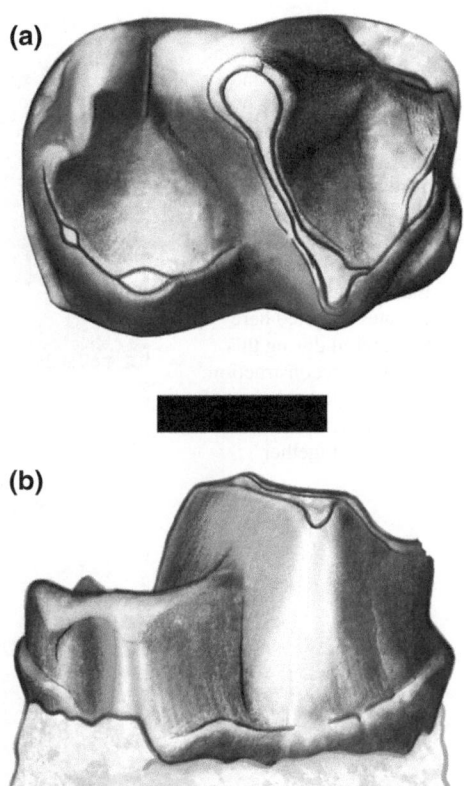

(a)

(b)

5.4 Paleogene Environmental Reconstruction of the Cucullaea I (Ypresian) and Submeseta (Priabonian) Biotas

Overall the knowledge and composition of the Paleogene Seymour Island biota allows environmental reconstruction for Ypresian (Fig. 5.25) and Priabonian (Fig. 5.26) settings and the association and interaction of the marine/coastal and terrestrial vertebrates in two different ages of the Eocene in West Antarctica.

5.5 Correlation of the *Cucullaea* I Terrestrial Fauna with Early Paleogene Patagonian Faunas

A taxonomic analysis of the terrestrial vertebrates from *Cucullaea* I Allomember of La Meseta Formation (Tables 4.5 and 5.6) reveals a modest taxonomic diversity that includes six avian and seven mammalian ordinal groups. This Early

Fig. 5.25 Reconstruction of the terrestrial, coastal and marine environments and vertebrate assemblage from the Early Eocene (Ypresian, 49–52 Ma) of Antarctic Peninsula, West Antarctica based on the paleontologic evidence from *Cucullaea* I Allomember of La Meseta Formation, Seymour Island. Vertebrates depicted here were analyzed during this study. In this reconstruction we are exercised a degree of artistic license to assemble these species together

Eocene vertebrate assemblage was probably even more diverse, because the documented diversity of the terrestrial vertebrates of the La Meseta Formation is, of course, minimal, being derived from a few sites and from small samples (less than 60 specimens).

Currently, Seymour Island land vertebrate-bearing localities are separated from the bulk of the Patagonian localities discussed herein by more than 20° of latitude (i.e., approximately 1,600 km). Considering that, on the basis of geologic and paleogeographic data (Lawver et al. 1992), the Drake Passage (about 1,000 km wide) started to open at about 36 Ma, then we can assume that, prior to that time, the distance between the Antarctic Peninsula and Patagonia ought to have been shorter (Fig. 1.1). The terrestrial mammals of Seymour Island greatly strengthen

Fig. 5.26 Reconstruction of the terrestrial, coastal and marine environments and vertebrate assemblage from the Late Eocene (Priabonian, 34 Ma) of Antarctic Peninsula, West Antarctica based on the paleontologic evidence from Submeseta Allomember of La Meseta Formation, Seymour Island. Vertebrates depicted here were analyzed during this study. In this reconstruction we are exercised a degree of artistic license to assemble these species together

the hypothesis that the *Cucullaea* I fauna had its origin in times that pre-date the early Eocene.

In general, the Paleogene terrestrial fauna from Seymour Island shows greatest faunal resemblance to older Paleogene faunas of Patagonia, i.e., Peligran and Riochican SALMAs (South American Land Mammal Ages) and Vacan "subage" (Reguero et al. 2002, Gelfo et al. 2009). The oldest faunas of the Paleogene of Patagonia range, from oldest to youngest), between the early late Paleocene Peligran SALMA (Selandian), late Paleocene "*Carodnia* faunal zone", latest Paleocene/early Eocene "*Kibenikhoria* faunal zone" (Itaboraian SALMA), early Eocene Riochican SALMA, and middle Eocene Vacan subSALMA. All these faunas are mainly known from the Golfo de San Jorge Basin, Chubut Province at

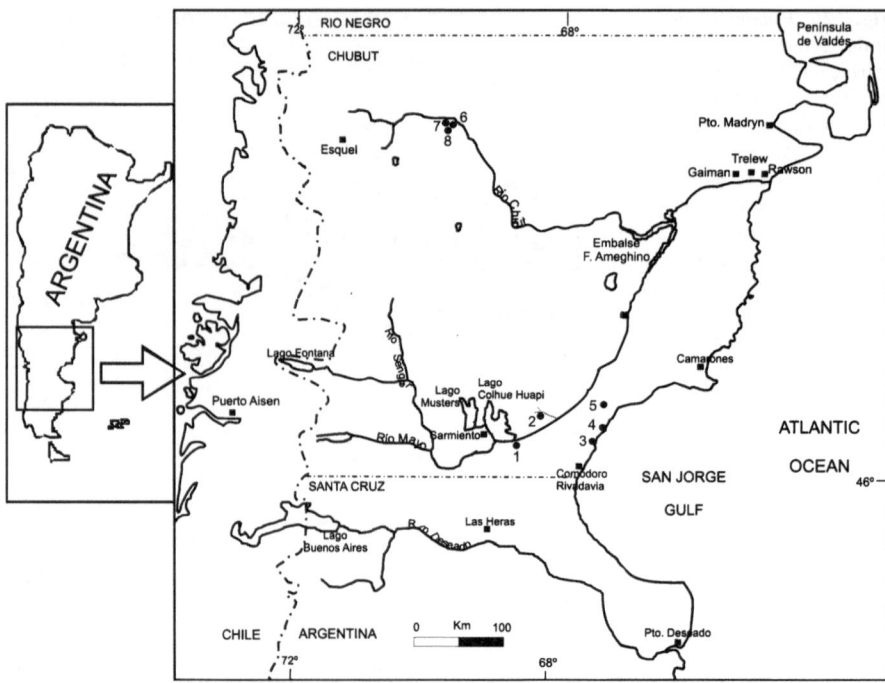

Fig. 5.27 Location map showing early Paleogene terrestrial mammal-bearing localities of Patagonia, Argentina discussed in the text: *1* Las Flores. *2* Cañadón Vaca. *3* Bajo Palangana. *4* Punta Peligro. *5* Cerro Redondo. *6* Grenier Farm. *7* Laguna Fría. *8* (modified from Gelfo 2006)

~45° S. A summary of the biostratigraphic, biochronologic and faunistic data of these Patagonian faunas is provided below.

The oldest South American Tertiary therian mammal from the earliest Danian-equivalent strata was recovered at Grenier Farm (Lefipán Fm., Danian, ~65 Ma), near Paso del Sapo, northwestern Chubut province, Argentina (Fig. 5.27) (Goin et al. 2006b). The specimen is an isolated lower molar most likely referable to a polydolopimorphian marsupial, a group known from the Late Cretaceous of North America as well as from later Paleocene and Eocene deposits in South America.

The Peligran SALMA was recognized on the basis of the mammalian association recovered from levels of the "Banco Negro Inferior" (Salamanca Formation, Hansen Member, ~61 Ma, early Selandian (Fig. 5.28) at Punta Peligro, Chubut (Fig. 5.27). It contains only six mammal taxa. Sudamericid gondwanatheres and derorhynchid marsupials are known from these horizons. As mentioned above, sudamericid gondwanatheres have recently been documented from Antarctica, albeit from younger, Eocene deposits (Reguero et al. 2002). These gondwanatheres, reported from the Eocene La Meseta Formation, Seymour Island, were considered by Goin et al. (2006a) to be closely related to *Sudamerica ameghinoi* (Sudamericidae) in their hypsodont crown morphology. A monotreme Ornithorhynchidae, represented by *Monotrematum sudamericanum* is

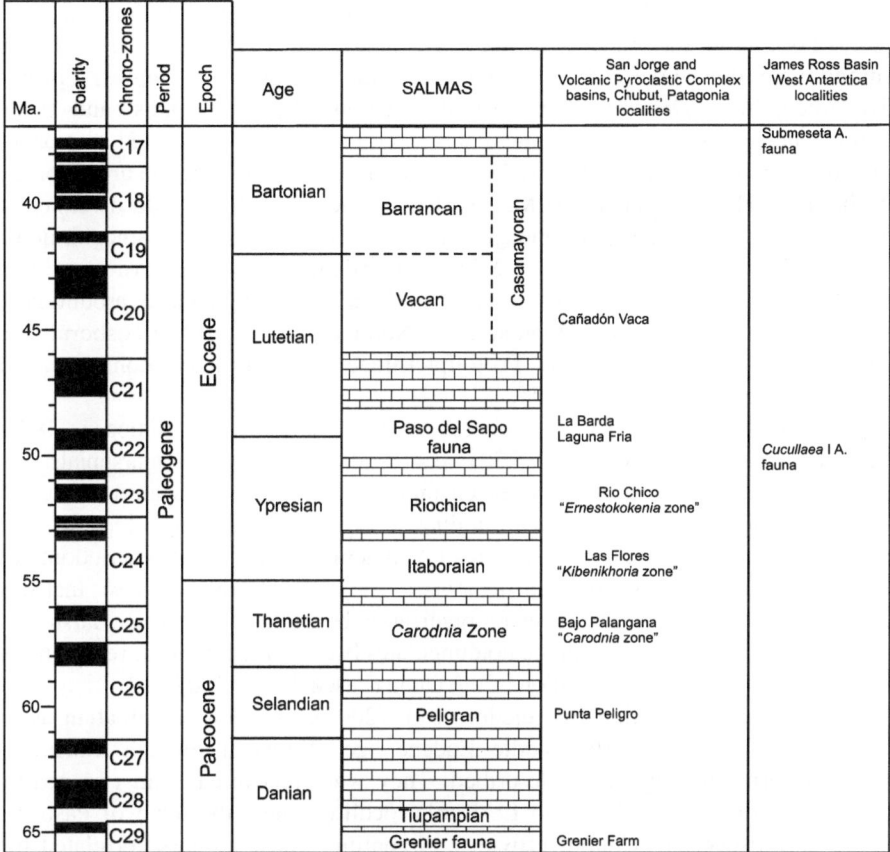

Fig. 5.28 Distribution of the principal Paleogene localities of the Golfo de San Jorge and "Volcanic-Pyroclastic Complex" basins, Chubut Province, Patagonia, Argentina (Gelfo et al. 2009) and the James Ross Basin, West Antarctica. Time scale according to Gradstein et al. (2004) and temperature curve according to Zachos et al. (2001)

also represented in the Punta Peligro fauna. The fauna of Punta Peligro is characterized by: (a) the presence of relictual Mesozoic non-therian mammals such as Gondwanatheria (Pascual et al. 1999) and Dryolestoidea (Gelfo and Pascual 2001); (b) exclusive Gondwanan lineages such as monotremes (Pascual et al. 1992); (c) the presence of derived native ungulates such as the notonychopid litopterns; and (d) the striking absence of Notoungulata.

Simpson's "*Carodnia* faunal zone" (~57 Ma, Thanetian) is poorly represented and studied, and includes a?borhyaenid indet, the polydolopid *Seumadia yapa*, the proterotheriid litoptern *Wainka tshotse*, and the pyrothere *Carodnia feruglioi*. Taxa from this horizon are apparently restricted to the Peñas Coloradas Formation in the San Jorge Basin and they appear to represent a new biochronologic unit, however, present evidence is meager and is not sufficient to warrant erecting a new SALMA (Bond et al. 1995).

Simpson's "*Kibenikhoria* faunal zone" (~58.0 Ma, Thanetian/Ypresian, Las Flores Formation, Itaboraian SALMA sensu Bond et al. 1995, Fig. 5.28) includes, among many other taxa, two polydolopines, *Epidolops* and *?Polydolops*; a primitive ?didelphoid, *Derorhynchus*; several protodidelphid marsupials; and seven families placed in four orders of ungulates. Astrapotheria is represented by a primitive trigonostylopid genus (*Shecenia*). Four families of Notoungulata are recorded in this age. The Itaboraian SALMA is also characterized by a wide diversity of mammals, several marsupial lineages (i.e., Borhyaenidae, Caroloameghiniidae, Derorhynchidae, Didelphidae, Microbiotheriidae, Bonapartheriidae, and Protodidelphidae); and Xenarthra (Dasypodidae), Xenungulata (Carodniidae), Astrapotheria (Trigonostylopidae), Notoungulata (Henricosborniidae, Oldfieldthomasiidae), Litopterna (Protolipternidae and Sparnotheriodontidae), and "archaic ungulates" Didolodontidae.

The Riochican SALMA (= "*Ernestokokenia* faunal zone" Simpson, 1935; Ypresian ~53.0–~54.0 Ma, Fig. 5.28) records three families of marsupials, the Polydolopidae being one of them. Seven families of Notoungulata are recorded in this age. This fauna shares three families (Polydolopidae, Prepidolopidae, and Sparnotheriodontidae) with the *Cucullaea* I Allomember fauna. Sparnotheriodontidae are represented by the genus *Victorlemoinea*. The available record shows that this fauna underwent notable taxonomic reorganization beginning at ~58 Ma (Marshall et al. 1997). During the Riochican the notoungulates became predominant, representing 44 % of the taxonomic composition of the fauna (Pascual et al. 2007).

In a recent contribution by Tejedor et al. (2009), two new mammalian associations of late early–early middle Eocene age (Ypresian-Lutetian transition) from western Patagonia were described. They were recovered from two nearby fossil sites, the Laguna Fría and La Barda localities, near the town of Paso del Sapo, in northwestern Chubut Province, Argentina. These authors correlated the Paso del Sapo faunas with that from the La Meseta Formation in the Antarctic Peninsula. This correlation was based on several shared taxa, such as the metatherian *Pauladelphys* (Derorhynchidae) and a litoptern close to *Notiolofos* (Sparnotheriodontidae). The cluster analysis of Gelfo et al. (2009) supported the close link between the Paso del Sapo and the La Meseta faunas, which were suggested as a distinct biochronological unit (Tejedor et al. 2009) between the Riochican SALMA and the Vacan subage of the Casamayoran SALMA (Fig. 5.29). These shared taxa suggest that the Antarctic connection may have been different and more favorable for the western Patagonian biotas than for those from eastern and central Patagonia (Woodburne and Zinsmeister, 1984; Woodburne, and Case, 1996). Perhaps the Paso del Sapo mammalian faunas represents a continental extension of the Weddellian Biogeographic Province into South America, as proposed by Zinsmeister (1979, 1982) on the basis of marine invertebrates (Tejedor et al. 2009). This paleobiogeographic relationship is also supported by the similar isotopic ages assigned to both sites. The mammal-bearing horizon within the *Cucullaea* I Allomember in the La Meseta Formation was dated to between 49 and 51 Ma using ^{87}Sr-^{86}Sr isotopic analysis (Ivany 2008), while those from Paso del Sapo span 47–52 Ma based on ^{40}K-^{40}Ar analysis (Tejedor et

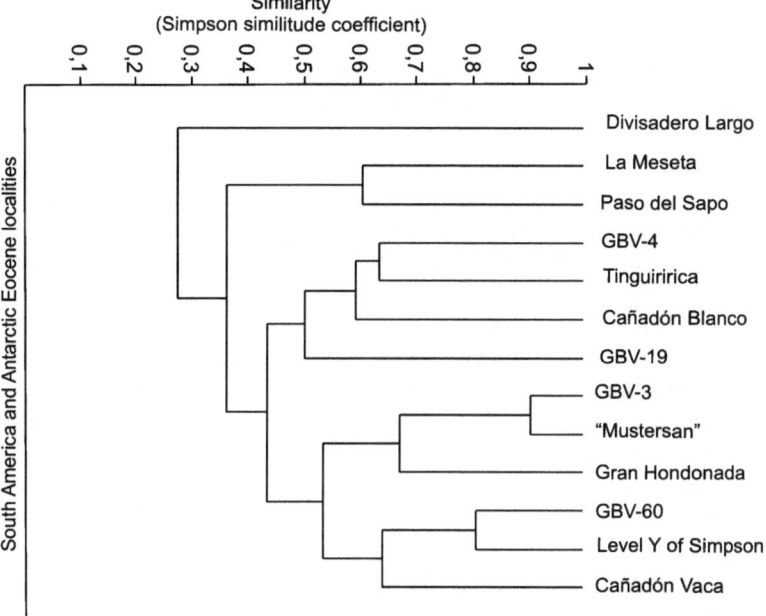

Fig. 5.29 Simpson's Coefficient of South American (Patagonia) and Antarctic (Seymour Island) Eocene localities mentioned in the text (after Gelfo et al. 2009)

al. 2009). These faunas likely fill the biochronologic gap between the Riochican and Casamayoran (Vacan subage) SALMAs (Tejedor et al. 2009).

Interestingly, the authors mentioned that the absence of Gondwanatheria in the Paso del Sapo faunas, and their presence in the Antarctic Peninsula, could have been due either to paleogeographic reasons or unsampling. The recent discovery of a gondwanatherian from La Barda adds evidence regarding the extensive radiation of the Gondwanatheria throughout the Southern Hemisphere and the persistence several lineages well after the Cretaceous/Paleogene boundary (Goin et al. 2012).

The Antarctic specimen seems to be more closely related to the Ferugliotheriidae, and thus shows a general, but not more specific affinity with the La Meseta specimen.

The Vacan fauna (Lutetian, Fig. 5.28) at Cañadón Vaca, Chubut (Sarmiento Formation) is lesser known than other Eocene faunas from Patagonia. It includes archaic notoungulate families (Henricosborniidae, Isotemnidae), and the relative primitiveness of this fauna document a great faunal difference from the subsequent Barrancan "subage" (Cifelli 1985).

The most unexpected circumstance in the La Meseta Formation is the apparent lack of notoungulates and other meridiungulate groups such as Condylarthra and non sparnotheriodontid Litopterna. Notoungulates were the most diverse (morphologically as well as taxonomically) and successful of the South American ungulate groups. One of the most important radiations of notoungulates occurred in Patagonia during the Late Paleocene-Early Eocene. As we suppose that no barrier to dispersal existed between Patagonia and the Antarctic Peninsula during the Paleocene, the absence of this group in Antarctica could be explained by suggesting that the La Meseta Formation is composed only of those taxa that were able to adapt to cooler conditions. However, the evidence of the presence of a high cordillera along the peninsular isthmus could have acted as a strong geographical barrier for the dispersal of terrestrial vertebrates in the Late Paleocene. Perhaps only vertebrates adapted to high altitudes were able to migrate further southward. Few ungulates have been recorded at Punta Peligro, Chubut (Peligran SALMA): the enigmatic *Peligrotherium tropicalis* (Dryolestida), the mioclaenid condylarths, and the notonychopid *Requisia vidmari* (?Litopterna). They document the existence of primitive meridiungulate lineages at this time. The first record of representatives of Notoungulata in Patagonia occurs in the late Paleocene '*Kibenikhoria* faunal zone' (Itaboraian SALMA sensu Bond et al. 1995) showing a discrete radiation of four families (Henricosborniidae, Isotemnidae, Interatheriidae, Oldfieldthomasiidae).

Interestingly, Bond (1999) remarks on the noteworthy difference in the geographic distribution between Notoungulata and Litopterna in the late Pleistocene (Lujanian) in Patagonia; whereas the litoptern *Macrauchenia patachonica* has a wide range of distribution southward (Santa Cruz Province) in this age, the notoungulate *Toxodon* reached only Bahia Blanca (38° 45'S). This fact suggests that some factor (geographic or environmental) affected the dispersal of notoungulates to southern latitudes during the Pleistocene.

If the notoungulates migrated southward into the Antarctic Peninsula during the late Paleocene, they presumably became extinct prior the deposition of the La Meseta Formation (late-early Eocene-late Eocene). In sum, the most plausible hypotheses for the absence of Notoungulata, and other groups, in West Antarctica are (1) the record of this group is taphonomically biased, or (2) they could have passed into West Antarctica during the latest part of the Paleocene when the environmental conditions were warmer, and then became extinct at the onset of the climatic deterioration (Early Eocene), or (3) the topography of the Antarctic Peninsula cordillera prevented the dispersal of this group into West Antarctica. Based on the evidence presented above the more plausible hypotheses is the second.

Seymour Island and the surrounding region (Antarctic Peninsula) started its faunal isolation from South America since the Late Paleocene and this might suggest that geographic isolation by a physical barrier like a seaway would be among the possible hypotheses available to explain the extinctions and the endemism of the fauna. Although a physical barrier (seaway) might not have developed until the end of the Eocene (opening of the Drake Passage), the cooling trend that began during the middle Eocene might have acted as an earlier barrier, discouraging new mammal immigrations. Therefore we suggest that regional cooling is the most reliable hypothesis to explain the extinctions, endemism, and relict character of the Eocene La Meseta Formation. Isolation that began through temperature decrease during the cooling trend from the middle Eocene onwards became physical with the development of the seaway between the Antarctic Peninsula and Patagonia at the end of the Eocene.

5.6 West Antarctic Terrestrial Biota and Its Intercontinental Relationships

In the Late Cretaceous of the James Ross Basin the majority of the non-avian dinosaurs (the megalosaur-like theropod, the ankylosaur, the dromaeosaurid, the ornithopods and the sauropod) are remnants of a cosmopolitan dinosaur fauna more typical of other areas (e.g., Campanian–Maastrichtian of Patagonia). Preliminary reports of both theropods and titanosaurians from the Late Cretaceous deposits in West Antarctica are suggestive in this regard. Furthermore, developing a more complete fossil record from other intervals and areas, e.g., post-Cenomanian of Africa) will provide information critical for reconstructing the biogeographical history of Cretaceous Gondwana. Thus, it is relevant to note that the recent discovery and/or re-appraisal of the non-avian dinosaur faunas of Australia and New Zealand has revealed typical Gondwanan elements, including the presence of lithostrotian titanosaur and theropods from the Early Cretaceous (Valanginian–Albian) (Smith et al. 2008; Hocknull et al. 2009). This firmly establishes the presence of these clades in East Gondwana and Patagonia before the Late Cretaceous.

The Cretaceous avian dinosaurs from West Antarctica recorded are rare, restricted to one charadriform and, with more certainty, a loon and a basal anseriform. The last two are flying birds whose life stories are linked mainly to water. If the assignment of the loon to *Neoageornis* is correct (it seems that it is), then we have another clear evidence of affinities of Antarctic fauna with that of southern Chile, although gaviids are clearly Holarctic. *Vegavis* gives the first strong evidence for a basal part of the extant avian radiation in the Cretaceous.

We agree with Woodburne and Case (1996) that the initial marsupial dispersal, from South America into and through West Antarctica and onward to Australia, must have occurred during the Late Cretaceous (?Maastrichtian). This process may have continued some time into the Paleocene (see below). This timing of events agrees with current evidence that supports a marine closure of the dispersal portal between West Antarctica and Australia at 64 Ma. The basal stocks of these clades, currently undocumented by fossils in the Late Cretaceous of South America, West Antarctica and Australia, is supported by stratigraphically calibrated phylogenies revealing long ghost lineages that extended into the Late Cretaceous (e.g., Meredith et al. 2008). In this regard, it is relevant to note that the recent discovery and/or re-appraisal of the non-avian dinosaur faunas of West Antarctica has revealed typical Gondwanan elements, including the presence of titanosaurian sauropods, ornithopods, and theropods from the Late Cretaceous (Coniacian/Campanian; Reguero et al. submitted).

Goin et al. (2007a) gave an alternative scenario on the problem of metatherian trans-Antarctic relationships. Following Morrone (e.g., 2002), they argued that, in considering Late Cretaceous-Paleogene times, it is misleading to refer Antarctica and the whole of South America to distinct and/or different biogeographic units. Northern and central South America seem to have belonged to one major biogeo-graphical unit (the Neotropical Region of the Holarctic Kingdom, Morrone 2002), while the southernmost tip of South America (the Andean Region, including Patagonia and the southern Andes), together with South Africa, Australasia, and Antarctica pertained to another (the Austral Kingdom, Morrone 2002). Goin et al. (2012) went further in arguing that the origins of the Austral biogeographic king-dom could be traced back to Late Triassic times.

The origin of microbiotherian marsupials and, probably, of other lineages of the Australidelphian radiation, was suggested to be restricted to the Andean bio-geographic region (Goin et al. 2007b). Considering that the Drake Passage was not completely open until well into Cenozoic times (probably, middle to late Eocene), it seems irrelevant to argue, for instance, a southern South American or an Antarctic origin for the Microbiotheria. On the contrary, the microbiotherian radiation should be regarded, as a whole, as an Austral Kingdom event (Goin et al. 2007b). The mutualistic relationships between the living microbiotheriid *Dromiciops gliroides* and a plant (*Tristerix*) related to the *Nothofagus* flora pro-vides an interesting clue on the possible trophic and biogeographic relationships of extinct Antarctic microbiotheriids. Other microbiotheriid associations, such as one proportionally rich assemblage from the early Neogene of Central Patagonia, have also been related to the nearby presence of a *Nothofagus* flora (Goin et al. 2007b).

Laza and Reguero (1990) proposed a southward extension of the fossil Patagonian herbivorous assemblage into Antarctica, on the record of two taxa of dung beetles, Phanaeinae and Dichotomiinae (Coleoptera: Polyphaga), in the *Cucullaea* I Allomember of the La Meseta Formation of Seymour Island. These beetles form balls of animal dung on which they feed and in which they lay their eggs (Waterhouse 1974). The association of these beetles and placental herbivo-rous in the Casamayoran and Mustersan rocks of Patagonia have been documented by Pascual (1980).

Another interesting case is that of the Polydolopidae (Marsupialia, Polydolopimorphia). Within South America, polydolopids never dispersed north of Patagonia and central Chile, as is also the case of almost all known microbioth-erians. On the contrary, they have been widely recorded in the Paleocene-Eocene of Patagonia, as well as in the La Meseta Fm in the Antarctic Peninsula. Because of this distribution pattern, they have been postulated as part of a marsupial radi-ation that was restricted to the Austral Kingdom (Chornogubsky et al. 2009). It should be noted, however, that they have not been recorded in any other region of the Southern Hemisphere (e.g., New Zealand, Australia). The distribution of the Polydolopidae argues in favor of a specific biogeographic terrestrial subregion (the Weddellian terrestrial province) within the Austral Kingdom. Here we wonder whether the Andean Region sensu Morrone (2002) should include, at least for the Late Cretaceous-late Eocene, not only Patagonia and the southern Andean Range but also Western Antarctica, New Zealand, and Tasmania as well. That is, a terrestrial extension of Zinsmeister's (1979, 1982) Weddellian Region, originally based on marine mollusks. It should be taken in account, however, that the fragmentation of some of the Weddellian land masses was well advanced by Late Cretaceous times. For instance, sea-floor formation at the New Caledonia Basin (i.e., the separation of New Zealand from East Antarctica) started already by 79–83 Ma (early Campanian; Sutherland 1999). Therefore, such separation anticipates any possible dispersal of metatherians from South America to New Zealand. Summing up, the Weddellian pattern of terrestrial vertebrate distributions throughout the Late Cretaceous-paleogene implies a quite complex, puzzle-like scenario in which the timing of different clades, as well as the timing of sea-floor formation, would give a distinct biogeographic pattern for ach stage along this time lapse.

As stated above, we favor a Late Cretaceous (?Maastrichtian) age for the initial dispersal of metatherians from South America to Australia through West Antarctica (see Woodburne and Case 1996). This may not have been the case for other mammalian groups. Eocene Antarctic sparnotheriodontids indicate that the oldest divergence time of the sparnotheriodontid clade from other basal litopterns is constrained by *Victorlemoinea* and thus lies in the Thanetian (Late Paleocene) some 4–5 Ma earlier than the known occurrence of the Seymour Island form (*Notiolofos*). Thus, the Antarctic astrapothere *Antarctodon* supports a Late Paleocene age for the dispersal to West Antarctica. This inferred time may have been already too late for the dispersal of South American native ungulates into other Gondwanian land masses (e.g., Australia, New Zealand).

Summing up, we conclude that (1) throughout the Late Cretaceous there was continuous causeway through West Antarctica and associated land bridges between South America and Australia; (2) small- to large-sized, obligate terrestrial forms (e.g., opossum-like marsupials, monotremes, ratite birds) gained broad distribution across these Gondwanan land masses prior to fragmentation and were isolated on Australia (marsupials) before the end of the Late Cretaceous; (3) based on phylogeny and calibrated stratigraphy of basal astrapotheres, the Weddellian land bridge between southern South America and West Antarctica may have been functional until the Thanetian (Late Paleocene).

References

Antoine, P-O, Marivaux L, Croft DA, Billet G, Ganerød M, Jaramillo C, Martin T, Orliac MJ, Tejada J, Altamirano AJ, Duranthon F, Fanjat G, Rousse S, Salas Gismondi R (2011) Middle Eocene rodents from Peruvian Amazonia reveal the pattern and timing of caviomorph origins and biogeography. Proc R Soc B published online 12 October 2011; electronic suppl material: http://dx.doi.org/ 10.1098/rspb.2011.1732
Archangelsky S (1972) Esporas de la Formación Rio Turbio (Eoceno). Revista del Museo de La Plata, (nueva serie): Paleontologia 39:65–100
Askin RA (1989) Endemism and heterochroneity in the Late Cretaceous (Campanian) to Paleocene palynofloras of Seymour Island, Antarctica: implications for origins, dispersal and paleoclimates of southern floras. In: Crame JA (ed) Origins and evolution of the Antarctic biota, vol 47. Geological Society Special Publication, London, pp 107–119

Askin RA (1990) Campanian to Paleocene spore and pollen assemblages of Seymour Island. Antarctica, Review of Palaeobotany and Palynology 65:105–113

Askin RA (1992) Late Cretaceous-Early Tertiary Antarctic outcrop evidence for past vegetation and climate. In: Kennett JP, Warnke DA (eds) The Antarctic paleoenvironment: a perspective on global change, vol 56. American Geophysical Union, Antarctic Research Series, Washington, pp 61–75

Askin RA (1997) Eocene–?earliest Oligocene terrestrial palynology of Seymour Island. In: Ricci CA (ed) The Antarctic Region: Geological Evolution and Processes. Terra Antarctica Publication, Siena, pp 993–996

Askin RA, Elliot DH, Stilwell JD, Zinsmeister WJ (1991) Stratigraphy and paleontology of Campanian and Eocene sediments, Cockburn Island, Antarctic Peninsula. Journal of South American Earth Science 4:99–117

Baldoni AM, Askin RA (1993) Palynology of the lower Lefi pan formation (upper cretaceous) of Barranca de Los Perros, Chubut Province, Argentina. Part II. Angiosperm pollen and discussion. Palynology 17:241–264

Barreda V, Palazzesi L (2007) Patagonian vegetation turnovers during the paleogene-early neogene: origin of arid-adapted floras. Bot Rev 73(1):31–50

Behrensmeyer AK, Damuth JD, Di Michele WA, Potts R, Sues H-D, Wing SL (1992) Terrestrial ecosystems through time. Evolutionary paleoecology of terrestrial plants and animals. The University of Chicago Press, Chicago, 578 pp

Berry EW (1937) Eogene plants from Rio Turbio, in the Territory of Santa cruz, Patagonia. Johns Hopkins Univ Stud Geol 12:91–98

Birkenmajer K (1981) Lithostratigraphy of the Point Hennequin Group (Miocene vulcanics and sediments) at King George Islands, Antarctica. Studia Geologica Polonica 72:59–73

Birkenmajer K (2001) Mesozoic and Cenozoic stratigraphic units in parts of the South Shetland Islands and northern Antarctic Peninsula (as used by the Polish Antarctic Programmes). Studia Geologica Polonica 118:5–188

Birkenmajer K, Zastawniak E (1989) Late Cretaceous-Early Tertiary floras of King George Island, West Antarctica: Their stratigraphic distribution and palaeoclimatic significance, origins and evolution of the Antarctic biota. Geol Soc Lond Spec Publ 147:227–240

Birkenmajer K, Delitala MC, Narebski W, Nicoletti M, Petrucciani C (1986) Geochronology and migration of Cretaceous through Tertiary plutonic centers, South Shetland Islands (West Antarctica): subduction and hot spot magmatism. Bulletin of the Polish Academy of Sciences, Earth Sciences 34:243–255

Birkenmajer K, Gazdzicki A, Krajewski KP, Przybycin A, Solecki A, Tatur A, Yoon HI (2005) First cenozoic glaciers in West Antarctica. Pol Polar Res 26:3–12

Bonaparte JF (1986) Sobre *Mesungulatum houssayi* y nuevos mamíferos cretácicos de Patagonia, Argentina. In: IV Congreso Argentino de Paleontología y Bioestratigrafía, vol 2, pp 48–61

Bond M (1999) Quaternary native ungulates of Southern South America. A synthesis. In: Rabassa J, Salemme M (eds) Quaternary of South America and Antarctic Peninsula. A.A. Balkema, Rotterdam, pp 177–205

Bond M, Pascual R, Reguero MA, Santillana SN, Marenssi SA (1990) Los primeros ungulados extinguidos sudamericanos de la Antártida. Ameghiniana 16:240

Bond M, Carlini AA, Goin FJ, Legarreta L, Ortiz Jaureguizar E, Pascual R, Uliana MA (1995) Episodes in South American land mammal evolution and sedimentation: testing their apparent concurrence in a Paleocene succession from Central Patagonia. In: VI Congreso Argentino de Paleontología y Bioestratigrafía Actas, pp 47–58

Bond M, Reguero MA, Vizcaíno SF, Marenssi SA (2006) A new "South American ungulate" (Mammalia: Litopterna) from the Eocene of the Antarctic Peninsula. In Francis JE, Pirrie D, Crame JA (eds) Cretaceous-tertiary high-latitude palaeoenvironments, James Ross Basin, Antarctica, vol 258. Geological Society of London, Special Publications, pp 163–176

Bond M, Reguero MA, Kramarz A, Moly JJ, Santillana SN, Marenssi SA (2008) Un Astrapotheria (Mammalia) del Eoceno de la Formación La Meseta, Isla Marambio (Seymour), Península Antártica. IV Simposio Latinoamericano sobre Investigaciones

Antárticas & VII Reunión Chilena de Investigación Antártica, Valparaíso 3 al 5 de septiembre de 2008, Resúmenes

Bond M, Reguero MA, Vizcaíno SF, Marenssi SA, Ortiz-Jaureguizar E (2009) *Notiolofos*, a replacement name for *Notolophus* Bond, Reguero, Vizcaíno and Marenssi, 2006, a preoccupied name. J Vertebr Paleontol 29:979

Bond M, Kramarz A, Macphee R, Reguero M (2011) A new astrapothere (Mammalia, Meridiungulata) from La Meseta Formation, Seymour (Marambio) Island, and a reassessment of previous records of Antarctic astrapotheres. Am Mus Novit 3718:16

Bowman VC, Francis JE, Riding JB, Hunter SJ, Haywood AM (2012) A latest cretaceous to earliest paleogene dinoflagellate cyst zonation from Antarctica, and implications for phytoprovincialism in the high southern latitudes. Rev Palaeobot Palynol 171:40–56

Boyde A, Fortelius M (1986) Development structure and function of rhinoceros enamel. Zool J Linn Soc 87:181–214

Buono MR, Fernández MS, Tambussi C, Mörs T, Reguero MA (2011) Un arqueoceto (Cetacea: Archaeoceti) del Eoceno temprano tardío de Isla Marambio (Formación La Meseta), Antártida. In: IV Congreso Latinoamericano de Paleontología de Vertebrados, 21 al 24 de Septiembre de 2011, San Juan, Argentina

Cambiaso A, Novas F, Lirio JM, Núñez H (2002) Un nuevo dinosaurio del Cretácico Superior de la Isla James Ross. Península Antártica, VIII Congreso Argentino de Paleontología y Bioestratigrafía, Corrientes, Resúmenes 61

Campbell KE Jr, Frailey CD, Romero-Pittman L (2004) The Paleogene Santa Rosa local fauna of Amazonian Perú: geographic and geologic setting. In: Campbell KE Jr (ed) The paleogene mammalian fauna of Santa Rosa, Amazonian Perú. Natural History Museum of Los Angeles County, Science Series 40, pp 3–14

Cantrill DJ, Nichols GJ (1996) Taxonomy and palaeoecology of early cretaceous (Late Albian) angiosperm leaves from Alexander Island, Antarctica. Rev Palaeobot Palynol 92:1–28

Cantrill DJ, Poole I (2002) Cretaceous patterns of floristic change in the Antarctic Peninsula. Geol Soc Lond Spec Publ 194:141–152

Cantrill DJ, Poole I (2005) Taxonomic turnover and abundance in cretaceous to tertiary wood floras of Antarctica: Implications for changes in forest ecology. Palaeogeogr Palaeoclimatol Palaeoecol 215:205–219

Carlini AA, Pascual R, Reguero MA, Scillato Yané GJ, Tonni EP, Vizcaíno SF (1990) The first paleogene land placental mammal from Antarctica: its paleoclimatic and paleobiogeographical bearings. In: IV international congress of systematic and evolutionary biology, Maryland, Abstracts, p 325

Carlini AA, Scillato-Yané GJ, Vizcaíno SF, Dozo MT (1992) Un singular Myrmecophagidae (Xenarthra, Vermilingua) de de Edad Colhuehuapense (Oligoceno tardio-Mioceno temprano) de Patagonia, Argentina. Ameghiniana 29:176

Case JA, Tambussi CP (1999) Maestrichtian record of neornithine birds in Antarctica: comment on a Late Cretaceous radiation of modern birds. J Vertebr Paleontol 19(3 supplement):37R

Case JA, Martin JE, Chaney DS, Reguero M, Marenssi SA, Santillana SM, Woodburne MO (2000) The first duck-billed dinosaur (Hadrosauridae) from Antarctica. Journal of Vertebrate Paleontology 20:612–614. doi:10.1671/0272-4634(2000)020[0612:TFDBDF]2.0.CO;2

Case JA, Martin JE, Chaney DS, Reguero M (2003) Late Cretaceous dinosaurs from the Antarctic Peninsula: remnant or immigrant fauna? J Vertebr Paleontol 23(3 Suppl):39A

Case JA, Reguero M, Martin JE, Cordes-Person A (2006) A cursorial bird from the Maastrichtian of Antarctica. Journal of Vertebrate Paleontology 26(3 supplement):48A

Case JA, Martin JE, Reguero MA (2007) A dromaeosaur from the Maastrichtian of James Ross Island and the Late Cretaceous Antarctic dinosaur fauna. U.S. Geological Survey and The National Academies; USGS OF-2007-1047, Short Research Paper, 083. doi:10.3133/of2007-1047.srp083

Cerda I, Paulina Carabajal A, Salgado L, Coria R, Moly J (2011) The first record of sauropod dinosaurs from Antarctica.71st Annual Meeting of the Society of Vertebrate Paleontology, Las Vega, USA. November, 2-5, 2011

Cerda IA, Paulina Carabajal A, Salgado L, Coria RA, Reguero MA, Tambussi CP, Moly JJ (2012) The first record of a sauropod dinosaur from Antarctica. Naturwissenschaften 99:83–87. doi:10.1007/s00114-011-0869-x

Chatterjee S (1989) The oldest Antarctic bird. J Vertebr Paleontol 9(3 Suppl):16A

Chatterjee S, Small BJ (1989) New plesiosaurs from the upper cretaceous of Antarctica. In: Crame JA (ed) Origin and evolution of the Antarctic biota, vol 47. Geological Society, Special Publication, London, pp 197–215

Chatterjee S, Martinioni D, Novas F, Mussel F, Templin R (2006) A new fossil loon from the Late Cretaceous of Antarctica and early radiation of foot-propelled diving birds. Journal of Vertebrate Paleontology, 26(3 supplement):49A

Chornogubsky L, Goin FJ, Reguero MA (2009) A reassessment of Antarctic polydolopid marsupials (Middle Eocene, La Meseta Formation). Antarct Sci 21:285–297

Cifelli RL (1985) Biostratigraphy of the Casamayoran, early Eocene, of Patagonia. Am Mus Novit 2820:1–26

Cione AL, Medina F (1987) A record of Notidanodon pectinatus (Chondrichtyes, Hexanchiformes) in the upper cretaceous of Antarctic Peninsula. Mesoz Res 1:79–88

Cione AL, Reguero MA, Acosta Hospitaleche C (2007) Did the continent and sea have different temperatures in the northern Antarctic Peninsula during the middle Eocene? Revista de la Asociación Geológica Argentina 62:586–596

Clarke JA, Tambussi CP, Noriega JI, Erickson GM, Ketcham RA (2005) Definitive evidence for the extant avian radiation in the cretaceous. Nature 433:305–308

Clarke JA, Tambussi CP, Noriega JI, Erickson GM, Ketcham RA (2006) Corrigendum to Definitive fossil evidence for the extant avian radiation in the Cretaceous. Nature 444:780

Coria RA, Tambussi C, Moly JJ, Santillana S, Reguero M (2007) Nuevos restos de Dinosauria del Cretácico de las islas James Ross y Marambio, Península Antárctica. VI Simposio Argentino y III Latinoamericano sobre Investigaciones Antárcticas, Dirección Nacional del Antárctico/Instituto Antártico Argentino – 10 al 14 de Septiembre de 2007, Instituto Antártico Argentino, Buenos Aires

Coria RA, Moly JJ, Reguero M, Santillana S (2008) Nuevos restos de Ornithopoda (Dinosauria, Ornithischia) de la Fm. Santa Marta, Isla James Ross, Antártida. Ameghiniana 45(Supl):25R

Cozzuol MA (1988) Comentarios sobre los Archaeoceti (Mammalia, Cetacea) de la isla Vicecomodoro Marambio, Antártida. In: Quiroga JC, Cione AL (eds) 5th Jornadas Argentinas de Paleontología Vertebrados, Abstracts, La Plata, p 32

Crame JA, Pirrie D, Riding JB, Thomson MRA (1991) Campanian-Maastrichtian (Cretaceous) stratigraphy of the James Ross Island area, Antarctica. J Geol Soc Lond 148:1125–1140. doi: 10.1144/gsjgs.148.6.1125

Crame JA, Francis JE, Cantrill DJ, Pirrie D (2004) Maastrichtian stratigraphy of Antarctica. Cretaceous Research 25:411–423

Cubitt G, Molloy L (1994) Wild New Zealand. The MIT Press, Cambridge, p 208

de la Fuente M, Novas FE, Isasi MP, Lirio JM, Nuñez HJ (2010) First cretaceous turtle from Antarctica. J Vertebr Paleontol 30:1275–1278

del Valle R, Medina F, Brandoni Z (1977) Nota preliminar sobre los hallazgos de reptiles fósiles marinos del suborden Plesiosauria en las islas James Ross y Vega, Antártida, vol 212. Contribuciones del Instituto Antártico Argentino, pp 1–13

Dettmann ME (1989) Antarctica: Cretaceous cradle of austral temperate rainforests? In: Crame JA (ed) Origins and evolution of the Antarctic biota. Geological Society Special Publications 47:89–105

Dettmann ME, Pocknall DT, Romero EJ, Zamaloa del MC (1990) Nothofagidites Erdtman ex Potonie, 1960; a catalogue of species with notes on the paleogeographic distribution of Nothofagus Bl. (southern beech). N Z Geol Surv Paleontol Bull 60:79

Dingle R, Lavelle M (1998) Antarctic Peninsula cryosphere: early oligocene (c. 30 Ma) initiation and a revised glacial chronology. J Geol Soc Lond 155:433–437

Dingle R, Marenssi S, Lavelle M (1998) High latitude Eocene climate deterioration: evidence from the northern Antarctic Peninsula. J S Am Earth Sci 11:571–579

Dolding PJD (1992) Palynology of the Marambio Group (Upper Cretaceous) of northern Humps Island. Antarctic Science 4(3):311–326

Dusén P (1907) Über die tertiäre flora der Magellansländer. Wissenschaftliche Ergebnisse der Schwedischen Expedition nach der Magellansländern, 1895–1897 vol 1(5), pp 241–248

Dusén P (1908) Über die tertiäre flora der Seymour-Insel. Wissenschaftliche Ergebnisse der Schwedischen Südpolar-Expedition 1901–1903, Lithographisches Institut des Generalstabs, Stokholm, Bd. 3, l. 3, 4 tf, p 127

Dutra TL (1997) Nothofagus leaf architecture, an old design to a new Gondwana: the use of the modern subtropical and temperate foliar character of the genus in paleoecology. II southern connection congress, 1997, Valdivia. Noticeros de Biologia 5:148–149

Dutra TL (2000a) Nothofagus no norte da Península Antártica (Ilha Rei George, ilhas Shetland do Sul) I. Cretáceo superior [Nothofagus in northern Antarctic Península (King George Island, South Shetland Islands. I. Late Cretaceous]. Revista da Universidade Guarulhos 5:102–105

Dutra TL (2000b) Nothofagus no norte da Península Antártica (Ilha Rei George, ilhas Shetland do Sul). II. Paleoceno superior—Eoceno Inferior [Nothofagus in northern Antarctic Península (King George Island, South Shetland Islands. II. Late paleocene-early eocene]. Revista da Universidade Guarulhos 5:131–136

Dutra TL (2004) Paleofloras da Antártica e sua relação com os eventos tectônicos e paleoclimáticos nas altas latitudes do sul [Antarctic paleofloras and its relation with tectonic and paleoclimatic events in high southern latitudes]. Revista Brasileira de Geociências 3(34):401–410

Dutra TL, Batten D (2000) The upper cretaceous flora from King George Island, an update of information and the paleobiogeographic value. Cretac Res 21(2–3):181–209

Dutton AL, Lohmann K, Zinsmeister WJ (2002) Stable isotope and minor element proxies for Eocene climate of Seymour Island, Antarctica. Paleoceanography 17(2):1–13

Fontes D, Dutra TL (2010) Paleogene imbricate-leaved podocarps from King George Island (Antarctica): assessing the geological context and botanical affinities. Revista Brasileira de Paleontologia 13(3):189–204

Fortelius M (1985) Ungulate cheek teeth: developmental, functional, and evolutionary interrelations. Acta Zool Fenn 180:1–76

Francis J, Hill R (1996) Fossil plants from the Pliocene Sirius Group. Transantarctic Mountains: evidence for climate from growth rings and fossil leaves, Palaios 11:389–396

Francis JE, Poole I (2002) Cretaceous and Early Tertiary climates of Antarctica: evidence from fossil wood. Palaeogeogr Palaeoclimatol Palaeoecol 182:47–64

Francis JE, Ashworth A, Cantrill DJ, Crame JA, Howe J, Stephens R, Tosolini A-M, Thorn V (2007) 100 million years of Antarctic climate evolution: evidence from fossil plants. In: Cooper AK, Barrett P, Stagg H, Storey B, Stump E, Wise W (eds) Antarctica: a keystone in a changing world. proceedings of the 10th international symposium on Antarctic earth sciences, Santa Barbara, California, pp 19–27

Frenguelli J (1941) Nuevos elementos florísticos del Magellaniano de Patagonia Austral. Notas del Museo de La Plata 6(30):173–202

Gandolfo, MA, Marenssi S A, Santillana SN (1998a) Flora y paleoclima de la Formación La Meseta (Eoceno medio), isla Marambio (Seymour), Antártida. In: Casadio S (ed) Paleógeno de América del Sur y de la Península Antártica, vol 5. Asociación Paleontológica Argentina, Publicación Especial, pp 155–162

Gandolfo, MA, Hoc P, Santillana S, Marenssi S (1998b) Una flor fósil morfológicamente afín a las Grossulariaceae (Orden Rosales) de la Formación La Meseta (Eoceno medio), Isla Marambio, Antártida. In: Casadio S (ed) Paleógeno de América del Sur y de la Península Antártica, vol 5. Asociación Paleontológica Argentina, Publicación Especial, pp 147–153

Gasparini Z, del Valle R, Goñi R (1984) Un elasmosáurido (Reptilia, Plesiosauria) del Cretácico Superior de la Antártida. Contribuciones del Instituto Antártico Argentino 305:1–24

Gasparini Z, Olivero E, Scasso R, Rinaldi C (1987) Un ankylosaurio (Reptila, Ornithischia) campaniano en el continente antártico, vol 1. In: Anais IV Congresso Brasileiro de Paleontologia, Rio de Janeiro, pp 131–141

Gayó E, Hinojosa LF, Villagrán C (2005) On the persistence of tropical paleofloras in central Chile during the early Eocene. Rev Palaeobot Palynol 137:41–50

Gelfo JN (2006) Los Didolodontidae (Mammalia, Ungulatomorpha) del Terciario sudamericano: sistemática, orígen y evolución. PhD dissertation, Facultad de Ciencias Naturales y Museo, UNLP, N884, pp 454

Gelfo JN, Pascual R (2001) *Peligrotherium tropicalis* (Mammalia, Dryolestida) from the early Paleocene of Patagonia, a survival from a Gondwanan radiation. Geodiversitas 23:369–379

Gelfo JN, Reguero MA, López GM, Carlini AA, Ciancio MR, Chornogubsky L, Bond M, Goin FJ, Tejedor M (2009) Eocene mammals and continental strata from patagonia and Antarctic Peninsula. Papers on Geology, Vertebrate Paleontology, and Biostratigraphy in Honor of Michael O. Woodburne. In: Albright III LB (ed) Museum of Northern Arizona Bulletin 64, Flagstaff, Arizona, pp 567–592

Goin FJ, Carlini A (1995) An Early Tertiary microbiotheriid marsupial from Antarctica. J Vertebr Paleontol 15:205–207

Goin FJ, Reguero MA (1993) Un "enigmático insectívoro" del Eoceno de Antártida. Ameghiniana 30:108

Goin FJ, Reguero MA, Vizcaíno SF (1995) Novedosos hallazgos de "comadrejas" (Marsupialia) del Eoceno medio de Antártida. In: III Jornadas de Comunicaciones sobre Investigaciones Antárticas, Buenos Aires, Resúmenes, pp 59–62

Goin FJ, Case JA, Woodburne MO, Vizcaíno SF, Reguero MA (1999) New discoveries of "opossum-like" marsupials from Antarctica (Seymour Island, Medial Eocene). J Mamm Evol 6(4):335–364

Goin F, Tejedor M, Bond M, López G, Reguero M (2000) Mamíferos Eocenos de Paso del Sapo. Chubut. Ameghiniana 37:25R

Goin FJ, Vieytes EC, Vucetich MG, Carlini AA, Bond M (2004) Enigmatic mammal from the Paleogene of Perú. In: Campbell Jr KE (ed) The paleogene mammalian fauna of Santa Rosa, Amazonian Perú. Natural History Museum of Los Angeles County, Science Series 40, Los Angeles, pp 145–153

Goin FJ, Reguero MA, Pascual R, von Koenigswald W, Woodburne MO, Case JA, Vieytes C, Marenssi SA, Vizcaíno SF (2006a) First gondwanatherian mammal from Antarctica. In: Francis JE, Pirrie D, Crame JA (eds) Cretaceous-tertiary high-latitude palaeoenvironments, James Ross Basin, Antarctica, vol 258. Geological Society of London, Special Publications, pp 135–144

Goin FJ, Pascual R, Tejedor MF, Gelfo JN, Woodburne MO, Case JA, Reguero MA, Bond M, López GM, Cione AL, Udrizar Sauthier D, Balarino L, Scasso RA, Medina FA, Ubaldón MC (2006b) The earliest tertiary therian mammal from South America. J Vertebr Paleontol 26(2):505–510

Goin FJ, Zimicz N, Reguero MA, Santillana SN, Marenssi SA, Moly JJ (2007a) New mammal from the Eocene of Antarctica, and the origins of the Microbiotheria. Revista de la Asociación Geológica Argentina 62:597–603

Goin FJ, Abello A, Bellosi E, Kay R, Madden R, Carlini AA (2007b) Los Metatheria sudamericanos de comienzos del Neógeno (Mioceno temprano, Edad-mamífero Colhuehuapense). Parte 1: Introducción, Didelphimorphia y Sparassodonta. Ameghiniana 44(1):29–71

Goin FJ, Tejedor MF, Chornogubky L, López GM, Gelfo JN, Bond M, Woodburne M, Gurovich Y, Reguero M (2012) Persistence of a mesozoic, non-therian mammalian lineage (Gondwanatheria) in the mid-paleogene of Patagonia. Naturwissenschaften

Gradstein FM, Ogg JG, Smith AG et al (2004) A geologic time scale 2004. Cambridge University Press, Cambridge

Grande L, Chatterjee S (1987) New cretaceous fish fossils from Seymour Island, Antarctic Peninsula. Palaeontology 30:829–837

Hammer WR, Hickerson WJ (1994) A crested theropod dinosaur from Antarctica. Science 264:828–830

Haomin L (1994) Early Tertiary Fossil Hill flora from Fildes Peninsula of King George Island, Antarctica. In: Shen Y (ed) Stratigraphy and palaeontology of Fildes Peninsula, King George Island. State Antarctic Committee Monograph 3, pp 133–171

Hill RS, Jordan GJ (1993) The evolutionary history of *Nothofagus*, (Nothofagaceae). Aust Syst Bot 6:111–126

Hinojosa LF, Armesto JJ, Villagrán C (2006) Are Chilean coastal forests pre-pleistocene relicts? Evidence from foliar physiognomy, palaeoclimate and phytogeography. J Biogeogr 33:331–341

Hocknull SA, White MA, Tischler RR, Cook AG, Calleja ND, Sloan T, Elliott DA (2009) New mid-Cretaceous (latest Albian) dinosaurs from Winton, Queensland, Australia. PLoS ONE 4:e6190. doi:10.1371/journal.pone.0006190

Hooker JJ (1992) An additional record of a placental mammal (Order Astrapotheria) from the Eocene of Western Antarctica. Antarct Sci 4:107–108

Hooker JJ, Milner AC, Sequeira SEK (1991) An ornithopod dinosaur from the Late Cretaceous of West Antarctica. Antarct Sci 3:331–332

Horovitz I, Ladèveze S, Argot C, Macrini TE, Martin T, Hooker JJ, Kurz C, de Muizon C, Sánchez-Villagra MR (2008) The anatomy of *Herpetotherium* cf. *fugax* Cope, 1873, a metatherian from the Oligocene of North America. Palaeontographica Abt A 284:109–141

Hünicken M (1955) Depósitos Neocretácicos y Terciarios del Extremo SW de Santa Cruz (Cuenca Carbonífera de Rio Turbio). Museo Argentino de Ciencias Naturales "Bernardino Rivadavia". Revista del Instituto Nacional del Investigacion de Ciencias Naturales, Ciencias Geologicas 4(1):164

Ivany LC, Lohmann KC, Hasiuk F, Blake DB, Glass A, Aronson RB, Moody RM (2008) Eocene climate record of a high southern latitude continental shelf: Seymour Island. Antarctica. Geological Society of America Bulletin 120(5):659–678

Janis CM (1976) The evolutionary strategy of the Equidae, and the origins of the rumen and cecal digestion. Evolution 30:757–774

Janis CM (1984) The use of fossil ungulate communities as indicators of climate and environment. In: Brenchley P (ed) Fossils and climates. Wiley, New York, pp 85–104

Keating JM (1992) Palynology of the Lachman Crags Member, Santa Marta Formation (Upper Cretaceous) of north-west James Ross Island. Antarct Sci 4(3):293–304

Krause DW, Prasad GVR, Koenigswald WV, Sahni A, Grine FE (1997) Cosmopolitanism among Gondwana Late Cretaceous mammals. Nature 390:504–507

Krause DW, Gottfried MD, O'Connor PM, Roberts EM (2003) A cretaceous mammal from Tanzania. Acta Palaeontol Polonica 48:321–330

Kriwet J (2003) First record of early cretaceous shark (Chondrichthyes, Neoselachii) from Antarctica. Antarct Sci 15:519–523

Kriwet K, Lirio JM, Nuñez HJ, Puceat E, Lécuyer C (2006) Late Cretaceous Antarctic fish diversity. In: Francis JE, Pirrie D, Crame JA (eds) Cretaceous-tertiary high-latitude palaeonvironments, James Ross Basin, Antarctica, vol 258. Geological Society of London, Special Publications pp 83–100

Lambrecht K (1929) Mesozoische und tertiäre Vogelreste aus Siebenbürgen. In: Csiki E (ed) X Congrés International de Zoologie. Stephaneum, Budapest, pp 1262–1275

Lawver LA, Gahagan LM, Coffin FM (1992) The development of palaeoseaway around Antarctica. In: Kennett JP, Warnke DA (eds) The Antarctic paleoenvironment: a perspective on global change vol 65. Antarctic Research Series, pp 7–30

Laza JH, Reguero MA (1990) Extensión faunística de la antigua región neotropical en la Península Antártica durante el Eoceno. Ameghiniana 26(3–4):245

Laza JH, Reguero MA (1993) Extensión faunística de la antigua región neotropical en la Península Antártica durante el Eoceno. Ameghiniana 26(3–4):245

MacPhee R, Reguero MA, Strgnac P, Nishida M, Jacobs P (2008) Out of Antarctica: Paleontological reconnaissance of Livingston Island (South Shetlands) and Seymour Island (James Ross Group). Journal of Vertebrate Paleontology, 28, Supplement to Number 3:110A

MacPhee RDE, Reguero MA (2010) Reinterpretation of the Middle Eocene record of Tardigrada (Pilosa, Edentata) from La Meseta Formation, Seymour Island, Antarctica. Am Mus Novit 3689:21

Marenssi SA, Santillana SN (1994) Unconformity-bounded units within the La Meseta Formation, Seymour Island, Antarctica: a preliminary approach. In: XXI polar symposium, Warszawa, Poland, Abstracts. Polish Academy of Sciences, Warsaw, pp 33–37

Marenssi SA, Lirio JM, Santillana SN (1992) The Upper Cretaceous of southern James Ross Island, Antarctica. In: Rinaldi CA (ed) Geología de la Isla James Ross. Antártida, Instituto Antártico Argentino, Buenos Aires, pp 89–99

Marenssi SA, Reguero MA, Santillana SN, Vizcaíno SF (1994) Eocene land mammals from Seymour Island, Antarctica: palaeobiogeographical implications. Antarct Sci 6:3–15

Margalef R (1983) Limnología. Barcelona, OMEGA, p 1009

Marshall LG, Sempere S, Butler RF (1997) Chronostratigraphy of the mammal-bearing Paleocene of South America. J S Am Earth Sci 10(1):49–70

Martin JE (2006) Biostratigraphy of the Mosasauridae (Reptilia) from the cretaceous of Antarctica. In: Francis JE, Pirrie D, Crame JA (eds) Cretaceous-tertiary high-latitude palaeonvironments, James Ross Basin, Antarctica, vol 258. Geological Society of London, Special Publications, pp 101–108

Martin JE, Crame JA (2006) Palaeobiological significance of high-latitude Late Cretaceous vertebrate fossils from James Ross Basin. Antarctica. In: Francis JE, Pirrie D, Crame JA (eds) Cretaceous-tertiary high-latitude palaeonvironments, James Ross Basin, Antarctica, vol 258. Geological Society of London, Special Publications, pp 109–124

Martin JE, Bell G Jr, Case JA, Chaney DS, Fernández MS, Gasparini Z, Reguero MA, Woodburne MO (2002) Late Cretaceous mosasaurs (Reptilia) from the Antarctic Peninsula. In: Gamble JA, Skinner DNB, Henrys S (eds) Antarctica and the close of the millenium, 8th international symposium on Antarctic earth sciences, vol 35. Bulletin of the Royal Society of New Zealand Bulletin, pp 293–299

Mayr G (2004) A partial skeleton of a new fossil loon (Aves, Gaviiformes) from the early Oligocene of Germany with preserved stomach content. Journal of Ornithology 145:281–286. doi:10.1007/s10336-004-0050-9

Meredith RW, Westerman M, Case JA, Springer MS (2008) A phylogeny and timescale for marsupial evolution based on sequences for five nuclear genes. Journal of Mammalian Evolution 15:1–36

Milner AC, Hooker JJ, Sequeira SEK (1992) An ornithopod dinosaur from the upper Cretaceous of the Antarctic Peninsula. Journal of Vertebrate Paleontology 12(3 supplement):44A

Mitchell ED (1989) A new cetacean from the late Eocene La Meseta Formation, Seymour Island, Antarctic Peninsula. Can J Fish Aquat Sci 46:2219–2235

Molnar RE, Angriman LA, Gasparini Z (1996) An Antarctic cretaceous theropod. Mem Qld Mus 39:669–674

Morrone JJ (2002) Biogeographical regions under track and cladistic scrutiny. J Biogeogr 29:149–152

Novas FE, Pol D (2005) New evidence on deinonychosaurian dinosaurs from the Late Cretaceous of Patagonia. Nature 433:858–861

Novas FE, Cambiasso AV, Lirio JM, Núñez HJ (2002) Paleobiogeografía de los dinosaurios polares de Gondwana. Ameghiniana 39 (Supl):15R

Olivero EB (1992) Asociaciones de ammonites de la Formación Santa Marta (Cretácico tardío), isla James Ross, Antártida. In: Rinaldi CA (ed) Geología de la Isla James Ross. Antártida, Instituto Antártico Argentino, Buenos Aires, pp 47–76

Olivero EB (2012) Sedimentary cycles, ammonite diversity and palaeoenvironmental changes in the Upper Cretaceous Marambio Group, Antarctica. Cretaceous Research 34:348–366. doi:10.1016/j.cretres.2011.11.015

Oliveira EV, Goin FJ (2011) A reassesment of bunodont metatherians from the Paleogene of Itaboraí (Brazil): Systematics and age of the Itaboraian SALMA. Revista Brasileira de Paleontologia 14(2):105–136

Olivero EB, Medina FA (2000) Patterns of Late Cretaceous ammonite biogeography in southern high latitudes: the Family Kossmaticeratidae in Antarctica. Cretaceous Research 21:269–279

Olivero E, Gasparini Z, Rinaldi C, Scasso R (1991) First record of dinosaurs in Antarctica (Upper Cretaceous, James Ross Island): paleogeographical implications. In: Thomson MRA,

Crame JA, Thomson JW (eds) Geological evolution of Antarctica. Cambridge University Press, Cambridge, pp 617–622

Olivero EB, Ponce JJ, Marsicano CA, Martinioni DR (2007) Depositional settings of the basal López de Bertodano Formation, Maastrichtian, Antarctica. Revista de la Asociación Geológica Argentina 62:521–529

Olson S (1992) *Neogaeornis wetzeli* Lambrecht, a Cretaceous loon from Chile (Aves: Gaviidae). Journal of Vertebrate Paleontology 12:122–124

Pankhurst RJ, Smellie J (1983) K-Ar Geochronology of the South Shetland Islands, Lesser Antarctica: apparent lateral migration of Jurassic to Quaternary island arc volcanism. Earth and Planetary Science Letters 66:214–222

Panti C (2011) Análisis paleoflorístico de la Formación Río Guillermo (Eoceno tardío–Oligoceno temprano?), Santa Cruz, Argentina. Ameghiniana 48(3):320–335

Panti C, Marenssi SA, Olivero EB (2008) Paleogene flora of the Sloggett Formation, Tierra del Fuego, Argentina. Ameghiniana 45(4):677–692

Pascual R (1980) Nuevos y singulares tipos ecológicos de marsupiales extinguidos de América del Sur (Paleoceno tardío o Eoceno temprano) del Noroeste Argentino. In: Bertels A, Romero E (eds) Actas II Congreso Argentino de Paleontología y Bioestratigrafía y I Congreso Latinoamericano de Paleontología, vol 2. Asociación Paleontológica Argentina, Buenos Aires, pp 151–173

Pascual R, Ortiz-Jaureguizar E (2007) The Gondwanan and South American episodes: two major and unrelated moments in the history of the South American mammals. J Mamm Evol 14:75–137

Pascual R, Archer M, Ortiz-Jaureguizar E, Prado JL, Godthelp H, Hand SH (1992) First discovery of monotremes in South America. Nature 356:704–705

Pascual R, Goin FJ, Krause DW, Ortiz-Jaureguizar E, Carlini AA (1999) The first gnathic remains of *Sudamerica*: implications for gondwanathere relationships. Journal of Vertebrate Paleontology 19:373–382

Partridge AD (2002) Quantitative palynological analysis of outcrop samples from the López de Bertodano Formation, James Ross Basin, northern Antarctic Peninsula. Biostrata Report 2002(25):1–52

Pirrie D, Crame JA, Riding JB (1991) Late Cretaceous stratigraphy and sedimentology of Cape Lamb, Vega Island, Antarctica. Cretaceous Research 12:227–258

Pirrie D, Marshall JD, Crame JA (1998) Marine high Mg calcite cements in *Teredolites* bored fossil wood: evidence for cool palaeoclimates in the Eocene La Meseta formation, Seymour Island, Antarctica. Palaios 13:276–286

Pole MS (1994) The New Zealand flora—entirely long-distance dispersal? J Biogeogr 21:625–635

Poole I, Hunt RJ, Cantrill D (2001) A fossil wood flora from King George Island: ecological implications for an Antarctic Eocene vegetation. Ann Bot 88:33–54

Poole I, Cantrill D, Utescher T (2005) A multi-proxy approach to determine Antarctic terrestrial palaeoclimate during the Late Cretaceous and Early Tertiary. Palaeogeogr Palaeoclimatol Palaeoecol 222:95–121

Prasad GVR, Verma O, Sahni A, Krause DW, Khosla A, Parmar V (2007) A new Late Cretaceous gondwanatherian mammal from central India. Proc Indian Natl Sci Acad 73:17–24

Premoli AC (1996) Leaf architecture of South American *Nothofagus* (Nothofagaceae) using traditional and new methods in morphometrics. Bot J Linn Soc 121:25–40

Premoli AC, Mathiasen P, Kitzberger T (2010) Southern-most *Nothofagus* trees enduring ice ages: Genetic evidence and ecological niche retrodiction reveal high latitude (54°S) glacial refugia. Palaeogeogr Palaeoclimatol Palaeoecol 298(3–4):247–256

Reguero MA, Gasparini Z (2007) Late Cretaceous-Early Tertiary marine and terrestrial vertebrates from James Ross Basin, Antarctic Peninsula: a review. In: Rabassa J, Borla ML (eds) Antarctic Peninsula and Tierra del Fuego: 100 years of Swedish-Argentine scientific cooperation at the end of the world. Taylor Francis, London, pp 55–76

Reguero MA, Marenssi SA (2010) Paleogene climatic and biotic events in the terrestrial record of the Antarctic Peninsula: an overview. In: Madden R, Carlini AA, Vucetich MG, Kay R (eds) The paleontology of Gran Barranca: evolution and environmental change through the Middle Cenozoic of Patagonia. Cambridge University Press, Cambridge, pp 383–397

Reguero MA, Vizcaíno SF, Goin FJ, Marenssi SA, Santillana SN (1998) Eocene high-latitude terrestrial vertebrates from Antarctica as biogeographic evidence. In: Casadio S (ed) Paleógeno de América del Sur y de la Península Antártica, vol 5. Asociación Paleontológica Argentina, Publicación Especial, pp 185–198

Reguero MA, Marenssi SA, Santillana SN (2002) Antarctic Peninsula and Patagonia Paleogene terrestrial environments: biotic and biogeographic relationships. Palaeogeogr Palaeoclimatol Palaeoecol 179:189–210

Rich T, Vickers-Rich P, Fernández M, Santillana S (1999) A probable hadrosaur from Seymour Island, Antarctica Peninsula. In: Tomida Y, Rich T, Vickers-Rich P (eds) Proceedings of the second Gondwana dinosaur symposium. National Science Museum, Tokyo, pp 219–222

Richter M, Ward DJ (1990) Fish remains from Santa Marta formation (Late Cretaceous) of James Ross Island, Antarctica. Antarct Sci 2:67–76

Romero E (1986) Paleogene phytogeography and climatology of South America. Ann Missouri Bot Garden 73(2):449–461

Romero EJ, Zamaloa MC (1985) Polen de angiospermas de la Formación Rio Turbio (Eoceno), Provincia de Santa Cruz, Republica Argentina. Buenos Aires, Ameghiniana 22(1–2):43–50

Rougier GW, Chornogubsky L, Casadío S, Paéz Arango N, Giallombardo A (2009) Mammals from the Allen formation, Late Cretaceous, Argentina. Cretac Res 30:223–238

Sadler P (1988) Geometry and stratification of uppermost cretaceous and paleogene units of Seymour Island, northern Antarctic Peninsula. In: Feldmann RM, Woodburne MO (eds) Geology and paleontology of Seymour Island, Antarctic Peninsula, vol 169. Geological Society of America, Memoir, pp 303–320

Salgado L, Gasparini Z (2006) Reappraisal of an ankylosaurian dinosaur from the Upper Cretaceous of James Ross Island (Antarctica). Geodiversitas 28:119–135

Sampson SD, Witmer LM, Forster CA, Krause DW, Connor PO, Dodson P, Ravoavy F (1998) Predatory dinosaur remains from Madagascar: implications for the Cretaceous biogeography of Gondwana. Science 280:1048–1051

dos Santos PR, Rocha-Campos AC, Tompette R, Uhlein A, Gipp M, Simões JC (1990) Review of Tertiary Glaciation in King George Island, West Antarctica: Preliminary results. Pesquisa Antártica Brasileira 2(1):87–99

Scillato Yané GJ, Pascual R (1984) Un peculiar Paratheria, Edentata (Mammalia) del Paleoceno medio de Patagonia. 1° Jornadas Argentinas de Paleontología de Vertebrados, Resúmenes 15. La Plata

Scillato-Yané G, Pascual R (1985) Un peculiar Xenarthra del Paleoceno Medio de Patagonia (Argentina). Su importancia en la sistemática de los Paratheria. Ameghiniana 21:173–176

Shen Y (1994) Cretaceous to Paleogene biostratigraphy in Antarctic peninsula and its significance in the reconstruction of Gondwanaland. In: Shen Y (ed) Stratigraphy and Palaeontology of Fildes Peninsula, King George Island, Antarctica. Monograph, Science Press, China 3:329–348

Sigé B, Sempere T, Butler RF, Marshall LG, Crochet J-Y (2004) Age and stratigraphic reassessment of the fossil-bearing Laguna Umayo red mudstone unit, SE Peru, from regional stratigraphy, fossil record, and paleomagnetism. Geobios 37:771–794

Simpson GG (1935) Occurrence and relationships of the Rio Chico Fauna of Patagonia. Am Mus Novit 818:1–21

Smith ND, Pol D (2007) Anatomy of a basal sauropodomorph dinosaur from the Early Jurassic Hanson Formation of Antarctica. Acta Palaeontol Polonica 52:657–674

Smith ND, Makovicky PJ, Agnolin FL, Ezcurra MD, Pais DF, Salisbury SW (2008) A Megaraptor-like theropod (Dinosauria: Tetanurae) in Australia: support for faunal exchange across eastern and western Gondwana in the mid-cretaceous. Proc R Soc B Biol Sci 275:2085–2093

Soliani E Jr, Bonhomme MG (1994) New evidence for Cenozoic resetting of K-Ar ages in volcanic rocks of the northern portion of the Admiralty bay, King George Island, Antarctica. Journal of South American Earth Sciences 7(1):85–94

Soliani JRE, Kawashita K, Fensterseifer HC, Hansen MAF, Troian FL (1988) K-Ar Ages of the Winkel Point Formation (Fildes Peninsula Group) and associated intrusions, King George Island, South Shetland Islands, Antarctica. Instit. Antártico Chileno, Série Científica 38:133–139

Stahl BJ, Chatterjee S (1999) A Late Cretaceous chimaerid (Chondrichthyes, Holocephali) from Seymour Island, Antarctica. Palaeontology 42:979–989

Stahl BJ, Chatterjee S (2002) A Late Cretaceous callorhynchid (Chondrichthyes, Holocephali) from Seymour Island, Antarctica. J Vertebr Paleontol 22:848–850

Stephens R (2008) Palaeoenvironmental and climatic significance of an Araucaria-dominated Eocene flora from Seymour Island, Antarctic. Ph D thesis, University of Leeds, Leeds, UK

Stilwell JD, Zinsmeister WJ (1992) Molluscan systematics and biostratigraphy: Lower Tertiary La Meseta Formation, Seymour Island, Antarctic Peninsula. Antarct Res Ser 55:1–152

Stuchlik L (1981) Tertiary pollen spectra from the Ezcurra Inlet Group of Admiralty Bay, King George Island (South Shetland Islands, Antarctica). Studia Geologica Polonica 72:109–130

Sutherland R (1999) Basement geology and tectonic development of the greater New Zealand region: an interpretation from regional magnetic data. Tectonophysics 308:341–362

Tambussi C, Acosta Hospitaleche C (2007) Antarctic birds (Neornithes) during the Cretaceous-Eocene times. Revista de la Asociación Geológica 62(4):604–617

Tambussi CP, Noriega JI, Santillana SN, Marenssi SA (1995) Falconid bird from the middle Eocene La Meseta Formation, Seymour Island, West Antarctica. J Vertebr Paleontol 15(Suppl to 3):55A

Tanai T (1986) Phytogeographic and phylogenetic history of the genus *Nothofagus* BL. (Fagaceae) in the southern hemisphere. J Fac Sci Hokkaido Univ Ser IV 21:505–582

Tejedor MF, Goin FJ, Gelfo JN, López GM, Bond M, Carlini AA, Scillato-Yané GJ, Woodburne MO, Chornogubsky L, Aragón E, Reguero MA, Czaplewski NJ, Vincon S, Martin GM, Ciancio MR (2009) New early Eocene mammalian fauna from western Patagonia, Argentina. Am Mus Novit 3638:1–43

Torres T (1990) Etude paleobotanique du tertiare de les Isles Roi George et Seymour, Antarctique. These de Doctorat, Laboratoire de Paléobotanique et Evolution de Végétaux de L'Université Claude Bernard Lyon, p 290

Torres T, Marenssi SA, Santillana SN (1994) Maderas fósiles de la isla Seymour, Formación La Meseta, Antártica. Serie Científica del INACH 44:17–38

Troedson AL, Riding JB (2002) Upper Oligocene to lowermost Miocene strata of King George Island, South Shetland Islands, Antarctica: stratigraphy, facies analysis, and implications for the glacial history of the Antarctic Peninsula. Journal of Sedimentary Research 72(4):510–523

Troncoso A, Romero EJ (1998) Evolución de las comunidades florísticas en el extremo sur de Sudamérica durante el Cenofítico. Monographs in Systematic Botany, Missouri Botanical Garden, vol 6. In Congreso Latinoamericano de Botánica Proceedings, pp 149–172

Vizcaíno SF, Reguero MA, Marenssi SA, Santillana SN (1997) New land mammal-bearing localities from the Eocene La Meseta Formation, Seymour Island, Antarctica. In: Ricci CA (ed) The Antarctic region: geological evolution and processes. Terra Antarctica Publication, Siena, pp 997–1000

Vizcaíno SF, Reguero MA, Goin FJ, Tambussi CP, Noriega JI (1998) Community structure of Eocene terrestrial vertebrates from Antarctic Peninsula. In: Casadio S (ed) Paleógeno de América del Sur y de la Peninsula Antártica, vol 5. Asociación Paleontológica Argentina, Publicación Especial, pp 177–183

von Koenigswald W, Goin F, Pascual R (1999) Hypsodonty and enamel microstructure in the Paleocene gondwanatherian mammal *Sudamerica ameghinoi*. Acta Palaeontol Polonica 44:263–300

Waterhouse DF (1974) The biological control of dung. Sci Am 230:100–109

Walker JD, Geissman JW (2009) (compilers) *Geologic Time Scale*. GSA Today: 60–61

Wilf P, Johnson KR, Cúneo NR, Smith ME, Singer BS, Gandolfo MA (2005) Eocene plant diversity at Laguna del Hunco and Río Pichileufú, Patagonia, Argentina. Am Nat 165:634–650

Wilson GP, Das Sarma DC, Anantharaman S (2007) Late Cretaceous sudamericid gondwanatherians from India with paleobiogeographic considerations of Gondwanan mammals. J Vertebr Paleontol 27:521–531

Wiman C (1905a) Vorfläufige Mitteilung über die alttertiaren Vertebraten der Seymourinsel. Bull Geol Inst Uppsala 6:247–253

Wiman C (1905b) Über die alttertiären Vertebraten der Seymourinsel.Wissenschaftliche Ergebnisse der Schwedischen Südpolar Expedition 1901–1903 3:1–37

Woodburne MO, Case JA (1996) Dispersal, vicariance and the Late Cretaceous to Early Tertiary land mammal biogeography from South America to Australia. J Mamm Evol 3:121–161

Woodburne MO, Zinsmeister WJ (1982) Fossil land mammal from Antarctica. Science 218:284–286

Woodburne MO, Zinsmeister WJ (1984) The first land mammal from Antarctica and its biogeographic implications. J Paleontol 54:913–948

Woodward AS (1908) On fossil fish-remains from snow hill and Seymour Islands. Wissenschaftliche Ergebnisse der Schwedischen Südpolar-Expedition 1901–1903 3:1–4

Yinxi W, Yanbin S (1994) Rb-Sr isotopic dating and trace element, REE geochemistry of Late Cretaceous volcanic rocks from King George Island, Antarctica. In: Yanbin S (ed), Stratigraphy and Palaeontology of Fildes Peninsula, King George Island, Antarctica, Beijing Science Press 3:109–131

Zachos J, Pagani M, Sloan L, Thomas E, Billups K (2001) Trends, rhythms, and aberrations in global climate 65 Ma to present. Science 292:686–693

Zachos JC, Dickens GR, Zeebe R (2008) An early Cenozoic perspective on greenhouse warming and carbon-cycle dynamics. Nature 451:279–283

Zastawniak E (1981) Tertiary leaf flora from the Point Hennequin Group of King George Island (South Shetland Island, Antarctica), preliminary report. Studia Geologica Polonica 72:97–108

Zastawniak E, Wrona R, Gazdzicki A, Birkenmajer K (1985) Plant remains from the top part of the point Hennequin group (Upper Oligocene), King George Island (South Shetland Islands, Antarctica). Studia Geologica Polonica 81:143–164

Zhian Z, Haomin L (1994b) Some gymnosperms from the Early Tertiary fossil hill flora in Fildes peninsula, King George Island, Antarctica. In: Shen Y (ed) Fildes stratigraphy and palaeontology of Peninsula, King George Island, Antarctica, vol 3. Monograph, Science Press, Beijing

Zhiyan Z, Haomin L (1994a) Some Late Cretaceous plants from King George Island, Antarctica. In: Shen Y (ed) Stratigraphy and palaeontology of Fildes Peninsula, King George Island, Antarctica. State Antarctic Committee, Monograph 3. Science Press, Beijing, pp 85–96, 3 pls [In Chinese, English abstract, including systematic descriptions]

Zinsmeister WJ (1979) Biogeographic significance of the Late Mesozoic and Early Tertiary molluscan faunas of Seymour Island (Antarctic Peninsula) to the final breakup of Gondwanaland. In: Gray J, Boucot A (eds) Historical Biogeography, plate tectonics and the changing environment, proceedings, 37th annual biological Colloquium and selected PAPERS. Oregon State University Press, Corvallis, pp 349–355

Zinsmeister WJ (1982) Late Cretaceous-Early Tertiary molluscan biogeography of southern Circum-Pacific. J Paleontol 56:84–102

Acknowledgments

Fieldwork was supported by the Instituto Antártico Argentino and Fuerza Aérea Argentina, which provided logistical support for our participation in the Antarctic fieldwork. Logistic support from the Fuerza Aérea Argentina officers and crew of helicopters and the Marambio Station is, as always, gratefully acknowledged.

We thank a number of colleagues for their assistance in the field; in particular we acknowledge the contributions made by Juan José Moly, Sergio Santillana, Rodolfo Coria, Claudia P. Tambussi, Ross MacPhee, Ignacio Cerda, Ariana Paulina Carabajal, Virginia Villamayor, Thomas Mörs and Jonas Hangstrom. We also have benefited from collaborative effort in the field (prospecting and picking) of Juan José Moly and Martín de los Reyes. To Thomas Mörs for providing access to the specimens of fossil penguins collected by the Swedish South Polar Expedition housed in the Naturhistoriska Riksmuseet of Stockholm.

We also thank Ross MacPhee (AMNH) and Javier N Gelfo (MLP) for critical review of an early draft of the manuscript. We thank Pablo Motta, Manuel Sosa and Marcela Tomeo for some of the art work and putting together the figures.

The authors acknowledge funding from the Instituto Antártico Argentino (PICTA 2004 and 2008), ANCYPT PICT 0365 and PICTO 0093 and Consejo Nacional de Investigaciones Científicas y Técnicas (CONICET, PIP 0361 to FJG).

Appendix
Systematic and List of the Specimens of the Weddellian Sphenisciformes from Cross Valley and La Meseta Formations, Seymour Island, Antarctic Peninsula

Systematic Paleontology
Sphenisciformes Sharpe 1891

Crossvallia unienwillia Tambussi, Reguero, Marenssi and Santillana 2005

> *Assigned materials.* MLP 00-1-10-1 (holotype, humerus, associated femur and tibiotarsus).
> *Ocurrence.* Cross Valley Formation, Late Paleocene (Tambussi et al. 2005).

Anthropornis nordenskjoeldi Wiman 1905

Assigned materials. MLP 93-X-1-4 (proximal epiphysis of humerus), MLP 82-IV-23-4 (proximal epiphysis of humerus), MLP 83-I-1-190 (proximal epiphysis of humerus), MLP 88-I-1-463 (proximal epiphysis of humerus), IB/P/B-0307 (distal humerus), IB/P/B-0478 (proximal humerus), IB/P/B-0711 (distal humerus), IB/P/B-0091 (proximal right humerus), IB/P/B-0092 (distal half of humerus), IB/P/B-0019 (complete humerus), IB/P/B-0463 (scapular portion of coracoid), IB/P/B-0837 (incomplete shaft coracoid), IB/P/B-0150 (complete ulna), IB/P/B-0613d (incomplete carpometacarpus), IB/P/B-0476 (incomplete distal femur), IB/P/B-0480 (incomplete distal femur), IB/P/B-0660 (incomplete distal femur), IB/P/B-0675 (distal femur), IB/P/B-0701 (femur without distal end), IB/P/B-0360 (distal end of tibiotarsus), IB/P/B-0501 (tibiotarsus without distal end), IB/P/B-0512 (shaft of tibiotarsus), IB/P/B-0536 (incomplete proximal end of tibiotarsus), IB/P/B-0636 (distal end of tibiotarsus), IB/P/B-0070 (fragmentary tarsometatarsus), IB/P/B-0287 (fragmentary tarsometatarsus), IB/P/B-0085 a and b (two fragments of tarsometatarsus), MLP 84-II-1-7 (fragmentary tarsometatarsus), MLP 83-V-20-50 (proximal end of tarsometatarsus), MLP 83-II-1-19 (incomplete proximal end of tarsometatarsus), IB/P/B-0575c (first phalanx of second digit), IB/P/B-0094a (incomplete quadrate), IB/P/B-0189 (fragment of mandible), IB/P/B-0684 (phalanx of digit III), IB/P/B-0250b (patella), IB/P/B-0823 (incomplete patella).

Occurrence. Submeseta Allomember but IB/P/B-0536 (Jadwiszczak 2006) from *Cucullaea* I Allomember (Myrcha et al. 2002, Tambussi et al. 2006, Jadwiszczak 2006) and Adelaide, (Australia) Oligocene (Jenkins 1974; Fordyce and Jones 1990).

Anthropornis grandis (Wiman 1905)

*Assigned materials.*MLP CX-60-25 (proximal epiphysis of humerus), MLP 83-V-30-5 (diaphysis of humerus), MLP 93-X-1-104 (complete humerus), IB/P/B-0179 (humerus without distal end), IB/P/B-0454 (fragmentary coracoid), IB/P/B-0064 (complete ulna), IB/P/B-0443 (ulna without distal end), IB/P/B-0483 (incomplete tarsometatarsus), MLP 83-V-20-84 (fragmentary tarsometatarsus), MLP 95-I-10-142 (incomplete tarsometatarsus), MLP 94-III-15-178 (incomplete tarsometatarsus), MLP 94-III-1-12 (fragmentary tarsometatarsus), MLP 86-V-30-19 (fragmentary tarsometatarsus), MLP 84-III-1-176 (fragmentary tarsometatarsus), MLP 84-II-1-66 (fragmentary tarsometatarsus), MLP 95-I-10-156 (fragmentary tarsometatarsus), MLP 93-X-1-149 (fragmentary tarsometatarsus).

Occurrence. Submeseta Allomember (Myrcha et al. 2002; Tambussi et al. 2006; Jadwiszczak 2006a) but IB/P/B-0454 from *Cucullaea* I Allomember.

Anthropornis sp.

Assigned materials. MLP 83-V-20-25 (proximal and distal epiphysis of humerus), MLP 83-V-20-28 (proximal epiphysis of humerus), MLP 93-X-1-105 (proximal epiphysis of humerus), MLP 83-V-20-402 (fragmentary diaphysis of humerus), MLP 93-X-1-4 (distal epiphysis of humerus), MLP 83-V-30-4 (proximal epiphysis of humerus), MLP 87-II-1-42 (proximal epiphysis of humerus), IB/P/B-0264c (proximal end of carpoetacarpus), IB/P/B-0620a (fragmentary carpometacarpus), IB/P/B-0716 (incomplete carpometacarpus).

Occurrence. Submeseta Allomember, but MLP 87-II-1-42 and IB/P/B-0716 that was found in Cucullaea I Allomember (Tambussi et al. 2006; Jadwiszczak 2006a).

Palaeeudyptes gunnari (Wiman 1905)

Assigned materials. MLP 82-IV-23-64 (diaphysis and proximal epiphysis of humerus), MLP 93-X-1-31 (complete humerus), MLP 82-IV-23-60 (proximal epiphysis of humerus), MLP 88-I-1-464 (proximal epiphysis of humerus), MLP 86-V-30-15 (proximal epiphysis of humerus), MLP 84-II-1-115 (proximal epiphysis of humerus), MLP 84-II-1-6 (proximal epiphysis of humerus), MLP 84-II-1-66 (proximal epiphysis of humerus), MLP 83-V-20-403 (proximal epiphysis of humerus), MLP 86-V-30-16 (proximal epiphysis of humerus), MLP 82-IV-23-59 (proximal epiphysis of humerus), MLP 84-II-1-41 (proximal epiphysis of humerus), MLP 83-V-20-51 (proximal epiphysis of humerus), MLP 95-I-10-226 (proximal epiphysis of humerus), MLP 93-X-1-30 (proximal epiphysis of humerus), MLP 91-II-4-262 (proximal epiphysis of humerus), MLP 88-I-1-469 (proximal epiphysis of humerus), IB/P/B- 0060 (proximal end of humerus), IB/P/B-0066 (fragmentary humerus), IB/P/B-0075 (proximal end of humerus), IB/P/B-0187 (proximal end of humerus), IB/P/B-0371 (proximal end of humerus), IB/P/B-0389 (proximal end of humerus), IB/P/B-0126 (proximal end of humerus), IB/P/B-0306 (complete humerus), IB/P/B-0373 (proximal end of humerus), IB/P/B-0451 (incomplete humerus), IB/P/B-0472 (complete humerus), IB/P/B-0573 (fragmentary humerus), IB/P/B-0105 (coracoid), IB/P/B-0151 (coracoid), IB/P/B-0613c (coracoid), IB/P/B-0175 (coracoid), IB/P/B-0136 (coracoid), IB/P/B-0345 (coracoid), IB/P/B-0083 (ulna), IB/P/B-0455 (fragmentary ulna), IB/P/B-0692 (proximal end of ulna), IB/P/B-0145 (fragmentary carpometacarpus), IB/P/B-0103 (femur), IB/P/B-0430 (femur), IB/P/B-0159 (distal end of femur), IB/P/B-0504 (incomplete femur), IB/P/B-0655 (incomplete femur), IB/P/B-0699 (fragmentary femur), IB/P/B-0137b (distal end of tibiotarsus), IB/P/B-0248b (distal end of tibiotarsus), IB/P/B-0161a (distal end of tibiotarsus), IB/P/B-0164a (proximal end of tibiotarsus), IB/P/B-0256 (proximal end of tibiotarsus), IB/P/B-0663 (proximal end of tibiotarsus), IB/P/B-0654 (complete tibiotarsus), IB/P/B-0409 (third digit of the second phalanx), IB/P/B-0413 (third digit of first phalanx), IB/P/B-0901 (third digit of the first phalanx), IB/P/B-0589c (third digit of the second phalanx), MLP 91-II4-222 (complete tarsometatarsus), IB/P/B-0072 (almost complete tarsometatarsus), IB/P/B-0112 (almost complete tarsometatarsus), IB/P/B-0277 (almost complete tarsometatarsus), IB/P/B-0487 (almost complete tarsometatarsus), IB/P/B-0124 (incomplete tarsometatarsus), IB/P/B-0286 (incomplete tarsometatarsus), IB/P/B-0294 (incomplete tarsometatarsus), IB/P/B-0295 (incomplete tarsometatarsus), IB/P/B-0296 (incomplete tarsometatarsus), IB/P/B-0541a (incomplete tarsometatarsus), MLP 87-II-1-45 (incomplete tarsometatarsus), MLP 82-IV-23-6 (incomplete tarsometatarsus), MLP 94-III-15-16 (incomplete tarsometatarsus), MLP 82-IV-23-5 (incomplete tarsometatarsus), MLP 84-II-1-75 (incomplete tarsometatarsus), MLP 84-II-1-6 (incomplete tarsometatarsus), MLP 83-V-20-27 (incomplete tarsometatarsus), MLP 93-X-1-151 (incomplete tarsometatarsus), MLP 95-I-10-16 (incomplete tarsometatarsus), MLP 84-II-1-47 (incomplete tarsometatarsus), MLP 84-II-1-65 (incomplete tarsometatarsus), MLP 84-II-1-124 (incomplete tarsometatarsus), MLP 83-V-20-41 (incomplete tarsometatarsus), MLP 83-V-20-34 (incomplete tarsometatarsus), MLP 93-X-1-84 (incomplete tarsometatarsus), MLP 84-II-1-24 (incomplete tarsometatarsus), MLP 93-X-1-112 (incomplete tarsometatarsus), MLP 93-X-1-117 (incomplete tarsometatarsus).

Occurrence. Submeseta Allomember but MLP 91-II-4-262, IB/P/B-0533 and MLP 88-I-1-469 come from *Cucullaea* I Allomember (Myrcha et al. 2002; Jadwiszcak 2006a).

Palaeeudyptes klekowskii Myrcha, Tatur and del Valle 1990

Assigned materials. MLP CX-60-201 (complete humerus), MLP 93-X-1-172 (complete humerus), MLP 93-X-1-3 (incomplete humerus), MLP CX-60-223 (complete humerus), MLP 82-IV-23-2 (diaphysis and proximal epiphysis of humerus), MLP 84-II-1-11 (diaphysis and proximal epiphysis of humerus), MLP 95-I-10-149 (diaphysis and proximal epiphysis of humerus), MLP 83-V-30-7 (diaphysis), MLP 83-V-30-3 (diaphysis and proximal epiphysis of humerus), MLP 82-IV-23-3 (proximal epiphysis of humerus), MLP 83-V-30-14 (proximal epiphysis of humerus), MLP 82-IV-23-1 (diaphysis and proximal epiphysis of humerus), MLP 83-V-20-30 (proximal epiphysis of humerus), MLP 84-II-1-2 (diaphysis and distal epiphysis of humerus), MLP CX-60-232 (diaphysis of humerus), MLP 84-II-1-12a (distal epiphysis of humerus), MLP 91-II-4-227 (distal epiphysis of humerus), MLP 93-X-1-174 (distal epiphysis of humerus), MLP 94-III-15-175 (complete humerus of humerus), MLP 95-I-10-217 (distal epiphysis of humerus), MLP 87-II-1-44 (distal epiphysis of humerus), IB/P/B-0141 (complete humerus), IB/P/B-0571 (humerus with shaft damaged), IB/P/B-0578 (complete humerus), IB/P/B-0854 and IB/P/B-0857 (incomplete shaft and sternal end of coracoid- probably from the same bone), IB/P/B-0133 (ulna without distal end), IB/P/B-0135 (ulna without distal end), IB/P/B-0344 (ulna), IB/P/B-0685 (ulna), IB/P/B-0503 (ulna), IB/P/B-0506 (proximal end of ulna), IB/P/B-0331 (carpometacarpus), IB/P/B-0248c (proximal end of tibiotarsus), IB/P/B-0357 (fragmentary tibiotarsus), IB/P/B-0369 (proximal end of tibbiotarsus), IB/P/B-0626 (complete tibiotarsus), IB/P/B-0192a (first phalanx of second digit), IB/P/B-0065 (incomplete tarsometatarsus), IB/P/B-0061 (incomplete tarsometatarsus), IB/P/B-0081 (incomplete tarsometatarsus), IB/P/B-0093 (incomplete tarsometatarsus), IB/P/B-0101 (incomplete tarsometatarsus), IB/P/B-0142 (incomplete tarsometatarsus), IB/P/B-0077 (tarsometatarsus), IB/P/B-0276 (tarsometatarsus), IB/P/B-0281 (tarsometatarsus), IB/P/B-0285 (tarsometatarsus), IB/P/B-0486 (tarsometatarsus), IB/P/B-0545 (tarsometatarsus), IB/P/B-0546 (tarsometatarsus), MLP 93-X-1-63 (tarsometatarsus), MLP 93-X-1-6 (tarsometatarsus), MLP 84-II-1-5 (tarsometatarsus), MLP 84-II-1-76 (tarsometatarsus), MLP 93-X-1-106 (tarsometatarsus), MLP 93-X-1-108 (tarsometatarsus), MLP 84-II-1-49 (tarsometatarsus), MLP 93-III-15-4 (tarsometatarsus), MLP 78-X-26-18 (tarsometatarsus), MLP 93-III-15-18 (tarsometatarsus), MLP 93-X-1-65 (tarsometatarsus), MLP 83-V-30-15 (tarsometatarsus), MLP 83-V-30-17 (tarsometatarsus), MLP 93-X-1-142 (complete tarsometatarsus), MLP 84-II-1-78 (complete tarsometatarsus), MLP 94-III-15-20 (complete tarsometatarsus), IB/P/B-0485 (complete tarsometatarsus).

Occurrence. All specimens from Submeseta Allomember except IB/P/B-0485, MLP 94-III-15-20 and MLP 84-II-1-78 (Myrcha et al. 2002; Jadwiszcak 2006a).

Palaeeudyptes antarcticus Huxley 1859

Assigned materials. MLP 84-II-1-1 (humerus without the proximal epiphysis).
Occurrence. Submeseta Allomember (Tambussi et al. 2006) and Oamaru locality, Late Eocene- Late Oligocene, New Zealand (Fordyce and Jones 1990).

Palaeeudyptes sp.

Assigned materials. IB/P/B- 0104 (incomplete coracoid), IB/P/B-0171 (incomplete coracoid), IB/P/B-0224 (incomplete coracoid), IB/P/B-0237 (incomplete coracoid), IB/P/B-0452 (incomplete coracoid), IB/P/B-0460 (incomplete coracoid), IB/P/B- 0461 (incomplete coracoid), IB/P/B- 0464 (incomplete coracoid), IB/P/B-0465b (incomplete coracoid), IB/P/B-0520 (incomplete coracoid), IB/P/B- 0521 (incomplete coracoid), IB/P/B-0530 (incomplete coracoid), IB/P/B-0559 (incomplete coracoid), IB/P/B-0587e (incomplete coracoid), IB/P/B-608a (incomplete coracoid), IB/P/B- 0611b (incomplete coracoid), IB/P/B-0611c (incomplete coracoid), IB/P/B-0613b (incomplete coracoid), IB/P/B-0616 (incomplete coracoid), IB/P/B-0827 (incomplete coracoid), IB/P/B-0828 (incomplete coracoid), IB/P/B-0830 (incomplete coracoid), IB/P/B-0831 (incomplete coracoid), IB/P/B-0834 (incomplete coracoid), IB/P/B-0842 (incomplete coracoid), IB/P/B-0844 (incomplete coracoid), IB/P/B-0846 (incomplete coracoid), IB/P/B-0850 (incomplete coracoid), IB/P/B-0851 (incomplete coracoid), IB/P/B-0855 (incomplete coracoid), IB/P/B-0856 (incomplete coracoid), IB/P/B-0858 (incomplete coracoid), IB/P/B-0859 (incomplete coracoid), IB/P/B- 0860 (incomplete coracoid), IB/P/B-0861 (incomplete coracoid), IB/P/B- 0862 (incomplete coracoid), IB/P/B-0873 (incomplete coracoid), IB/P/B-0875 (incomplete coracoid), IB/P/B-0876 (incomplete coracoid), IB/P/B-0880 (incomplete coracoid), IB/P/B-0881 (incomplete coracoid), IB/P/B-0882 (incomplete coracoid), IB/P/B-0884 (incomplete coracoid), IB/P/B-0098 (incomplete humerus), IB/P/B-0379 (incomplete humerus), IB/P/B-0388 (incomplete humerus), IB/P/B-0390 (incomplete humerus), IB/P/B-0453 (incomplete humerus), IB/P/B-0700 (incomplete humerus), IB/P/B-0703 (incomplete humerus), IB/P/B-0719 (incomplete humerus), IB/P/B-0720 (incomplete humerus), IB/P/B-0737 (incomplete humerus), IB/P/B-0401 (incomplete tibiotarsus), IB/P/B-0634 (incomplete tibiotarsus), IB/P/B-0662 (incomplete tibiotarsus), IB/P/B-0537 (complete tibiotarsus), IB/P/B-0249b (first phalanx of second digit), IB/P/B-0651d (first phalanx of second digit), IB/P/B-0414 (first phalanx of fourth digit), IB/P/B-0896 (first phalanx of fourth digit), IB/P/B-0420 (first phalanx of second digit), IB/P/B-0424 (first phalanx of second digit), IB/P/B-0589d (first phalanx of second digit), IB/P/B-0895 (first phalanx of second digit), IB/P/B-0904 (first phalanx of second digit), IB/P/B-0904 (first phalanx of second digit), IB/P/B-0907 (first phalanx of second digit), IB/P/B-0913 (first phalanx of second digit), IB/P/B-0916 (first phalanx of second digit).

Ocurrence. Cucullaea I Allomember (Jadwiszcak 2006a).

Delphinornis larseni Wiman 1905

Assigned materials. MLP 93-X-1-147 (near complete humerus, lacks the distal end), MLP 93-X-1-146 (complete humerus), MLP 84-II-1-169 (diaphysis and fragmentary proximal epiphysis of humerus), MLP 93-X-1-21 (diaphysis of humerus), MLP 84-II-1-16 (diaphysis and fragmentary proximal epiphysis of humerus), MLP 93-X-1-32 (diaphysis and proximal epiphysis of humerus), MLP 93-X-1-144 (diaphysis and distal epiphysis of humerus), MLP 94-III-15-177 (near complete, lacks the proximal end of humerus), MLP 91-II-4-263 (proximal epiphysis of humerus), IB/P/B-0062 (complete tarsometatarsus), IB/P/B-0280 (incomplete tarsometatarsus), IB/P/B-0299 (incomplete tarsometatarsus), IB/P/B-0547 (incomplete tarsometatarsus), IB/P/B-0548 (incomplete tarsometatarsus), MLP 83-V-20-5 (complete tarsometatarsus), MLP 91-II-4-174 (almost complete tarsometatarsus), MLP 84-II-1-179 (incomplete tarsometatarsus), IB/P/B-0337 (distal end of tibiotarsus).

Ocurrence. Submeseta Allomember, but MLP 94-III-15-177 and MLP 91-II-4-263 come from the *Cucullaea* I Allomember (Myrcha et al. 2002; Jadwiszcak 2006a).

Delphinornis arctowskii Myrcha, Jadwiszczak, Tambussi, Noriega, Gazdzicki, Tatur and del Valle 2002

Assigned materials. IB/P/B-0115 (strongly eroded tarsometatarsus), IB/P/B-0266 (tibiotarsus without proximal end), IB/P/B-0500 (tibiotarsus without distal half), IB/P/B-0484 (complete tarsometatarsus), MLP 93-X-1-92 (incomplete tarsometatarsus).

Occurrence. Submeseta Allomember (Myrcha et al. 2002).

Delphinornis gracilis Myrcha, Jadwiszczak, Tambussi, Noriega, Gazdzicki, Tatur and del Valle 2002

Assigned materials. IB/P/B-0408 (fragmentary tibiotarsus).
Occurrence. Submeseta Allomember (Jadwiszcak 2006a).

Delphinornis cf. arctowskii Myrcha, Jadwiszczak, Tambussi, Noriega, Gazdzicki, Tatur and del Valle 2002

Assigned materials. MLP 93-X-1-70 (near complete humerus).
Occurrence. Submeseta Allomember.

Mesetaornis polaris Myrcha, Jadwiszczak, Tambussi, Noriega, Gazdzicki, Tatur and del Valle 2002

Assigned materials. IB/P/B-0278 (nearly complete tarsometatarsus).
Occurrence. Submeseta Allomember (Myrcha et al. 2002, Jadwiszcak 2006a).

Mesetaornis sp.

Assigned materials. IB/P/B-0279b (incomplete tarsometatarsus).
Occurrence. Submeseta Allomember (Jadwiszcak 2006a).

Marambiornis exilis Myrcha, Jadwiszczak, Tambussi, Noriega, Gazdzicki, Tatur and del Valle 2002.

Assigned materials. IB/P/B-0490 (complete tarsometatarsus), MLP 93-X-1-111 (complete tarsometatarsus).
Occurrence. Submeseta Allomember (Jadwiszcak 2006a).

Archaeospheniscus lopdelli Marples 1952

Assigned materials. MLP 94-III-15-17 (complete humerus), MLP 93-X-1-123 (proximal epiphysis of humerus), MLP 93-X-1-27 (proximal epiphysis of humerus), MLP 95-I-10-231 (diaphysis and distal epiphysis of humerus), MLP 95-I-10-236 (proximal epiphysis of humerus), MLP 84-II-1-110 (diaphysis and distal epiphysis of humerus), MLP 95-I-10-227 (diaphysis and proximal epiphysis of humerus), MLP 84-II-1-111 (diaphysis and proximal epiphysis of humerus), MLP 93-X-1-97 (diaphysis and distal epiphysis of humerus), MLP 95-I-10-233 (diaphysis and distal epiphysis of humerus).
Occurrence. Submeseta Allomember (Myrcha et al. 2002).

Archaeospheniscus wimani (Marples 1953)

Assigned materials. IB/P/B-0466 (incomplete coracoid), IB/P/B-0467 (incomplete coracoid), IB/P/B-0608b (incomplete coracoid), IB/P/B-0176 (incomplete humerus), IB/P/B-0641 (complete femur), IB/P/B- 0658 (shaft femur), IB/P/B-0687 (shaft femur), IB/P/B-0110 (tibiotarsus), IB/P/B-0137a (proximal end of tibiotarsus), IB/P/B-0218 (shaft of tibiotarsus), IB/P/B-0802 (shaft of tibiotarsus), IB/P/B-0796 (incomplete shaft of tibiotarsus), IB/P/B-0908 (first phalanx of third digit), IB/P/B-0284 (incomplete tasometatarsus), IB/P/B-0289 (incomplete tarsometatarsus), IB/P/B-0491 (incomplete tarsometatarsus), MLP 90-I-20-24 (complete tarsometatarsus), MLP 91-II-4-173 (incomplete tarsometatarsus).

Occurrence. Cucullaea I and Submeseta Allomembers (Myrcha et al. 2002).

Tonniornis mesetaensis Tambussi, Acosta Hospitaleche, Reguero and Marenssi 2006

Assigned materials. MLP 93-X-1-145 (holotype, complete humerus).
Ocurrence. Submeseta Allomember (Tambussi et al. 2006).

Tonniornis minimum Tambussi, Acosta Hospitaleche, Reguero and Marenssi 2006

Assigned materials. MLP 93-I-6-3 (holotype, complete humerus), MLP 93-X-1-22 (diaphysis and distal epiphysis of humerus).
Ocurrence. Submeseta Allomember (Tambussi et al. 2006).